John Alexander Henderson

Henderson & Hamlin's Lightning Calculator

John Alexander Henderson

Henderson & Hamlin's Lightning Calculator

ISBN/EAN: 9783337250522

Printed in Europe, USA, Canada, Australia, Japan

Cover: Foto ©berggeist007 / pixelio.de

More available books at **www.hansebooks.com**

HENDERSON & HAMLIN'S

LIGHTNING CALCULATOR;

CONTAINING

The Shortest, Simplest, and most Rapid Method of Computing Numbers, adapted to all kinds of Business, and within the Comprehension of every one having the slightest knowledge of Figures.

———————•———————

ENERGY IS THE PRICE OF SUCCESS.

———————•———————

"The methods of calculation, by Prof. J. A. HENDERSON, are invaluable to business men, and will prove a light in science to all coming generations."—A. J. WARNER, Pres. Elmira Commercial College.

"The above two methods are the finest known for lightning multiplication."—Prof. D. R. FORD, Female College, Elmira.

"I have examined Prof. J. A. HENDERSON'S new methods of calculation: They are remarkable for originality, and of great practical value. His methods of calculating Interest are peculiarly clear and comprehensive in their adaptation to all possible cases."—Rev. Dr. O. P. FITZGERALD

———————••———————

Address all orders for this Book to

Prof. J. A. HENDERSON,

SAN FRANCISCO, CAL.

SAN FRANCISCO:

A. L. BANCROFT & CO., PRINTERS AND LITHOGRAPHERS,

721 Market Street, San Francisco.

1872.

PREFACE.

It is better to know everything about something, than something about everything.

Early ideas are not usually true ideas, but need to be revised and re-revised. Right means straight, and wrong means crooked. And knowing that thought kindles at the fire of thought, we do not hesitate or offer any apology for presenting to the Public some new seed-thoughts, and right methods of operation in business calculations. The practical utility of this book is found in the brevity and conciseness of its rules. Particular attention is invited to the grand improvements in the subjects of computing time, all possible cases in Interest, Squaring and Multiplying Numbers, Dividing and Multiplying Fractions, and an infinite number of methods of Extracting Square and Cube Root.

ADDITION.

To be able to add two, three, or four columns of figures at once is deemed by many to be a Herculean task, and only to be accomplished by the gifted few; or, in other words, by mathematical prodigies. If we can succeed in dispelling this illusion, it will more than repay us ; and we feel very confident that we can, if the student will lay aside all prejudice, bearing steadily in mind that to become proficient in any new branch or principle, a little wholesome application is necessary. On the contrary, we cannot teach a student who takes no interest in the matter, one who will always be a drone in society. Such men have no need of this principle.

If two, three, or more columns can be carried up at a time, there must be some law or rule by which it is done. We have two principles of Addition ; one for adding short columns, and one for adding very long columns. They are much alike, differing only in detail. When one is thoroughly learned, it is very easy to learn the second. By a little attention to the following example, much time in future will be saved.

ADDITION OF SHORT COLUMNS OF FIGURES.

Addition is the basis of all numerical operations, and is used in all departments of business. To aid the business man in acquiring facility and accuracy in adding short columns of figures, the following method is presented as the best :—

```
 274
 346
 134
 342
 727
 329
─────
2152.
```

PROCESS.—Commence at the right hand column, add thus : 16, 22, 32; then carry the 3 tens to the second column; then add thus: 7, 14, 25; carry the 2 hundreds to the third column, and add the same way: 12, 16, 21.

In this way you name the sum of two figures at once, which is quite as easy as it is to add one figure at a time. Never permit yourself, *for once*, to add up a column in this manner: 9 and 7 are 16, and 2 are 18 and 4 are 22, and 6 are 28, and 4 are 32. It is just as easy to name the result of two figures at once, and four times as rapid.

The following method is recommended for the

ADDITION OF LONG COLUMNS OF FIGURES.

In the addition of long columns of figures, which frequently occur in books of accounts, in order to add them with certainty, and, at the same time, with ease and expedition, study well the following method, which practice will render familiar, easy, rapid, and certain.

THE EASY WAY TO ADD.

EXAMPLE 2—EXPLANATION.

Commence at 9 to add, and add as near 20 as possible, thus : $9+2+4+3=18$, place the 8 to the right of the 3, as in example ; commence at 6 to add $6+4+8=18$; place the 8 to the right of the 8, as in example ; commence at 6 to add $6+4+7=17$; place the 7 to the right of the 7, as in example; commence at 4 to add $4+9+3=16$; place the 6 to the right of the 3, as in example ; commence at 6 to add $6+4+7=17$; place

```
7⁷
4
6
3⁶
9
4
7⁷
4
6
8⁸
```

the 7 to the right of the 7 as in example; now, having arrived at the top of the column, we add 3 and figures in the new column, thus: $7+6+7+8+8=36$; place the right-hand figure of 36, which is a 6, under the original column, as in example, and add — the left-hand figure, which is a 3, to the number of figures in the new column; there are 5 figures in the new column, therefore $3+5=8$; prefix the 8 with the 6, under the original column, as in example; this makes 86, which is the sum of the column.

    4
    6
    3⁸
    4
    2
    9
    86

Remark 1.—If, upon arriving at the top of the column, there should be one, two or three figures whose sum will not equal 10, add them on to the sum of the figures of the new column, never placing an extra figure in the new column, unless it be an excess of units over ten.

Remark 2. — By this system of addition you can stop at any place in the column, where the sum of the figures will equal 10 or the excess of 10; but the addition will be more rapid by your adding as near 20 as possible, because you will save the forming of extra figures in your new column.

EXAMPLE—EXPLANATION.

$2+6+7=15$, drop 10, place the 5 to the right of the 7; $6+5+4=15$, drop 10, place the 5 to the right of the 4, as in example; $8+3+7=18$, drop 10, place the 8 to the right of the 7, as in example; now we have an extra figure, which is 4; add this 4 to the top figure of the new column, and this sum on the balance of the figures in the new

    5
    5
    4
    7⁸
    3
    8
    4⁵
    5

column, thus: $4+8+5+5=22$; place the right-hand figure of 22 under the original column, as in example, and add the left-hand figure of 22 to the number of figures in the new column, which are three, thus: $2+3=5$; prefix this 5 to the figure 2, under the original column; this makes 52, which is the sum of the column.

    6
    7⁵
    6
    2
    —
    52

RULE.—*For adding two or more columns, commence at the right-hand, or units' column; proceed in the same manner as in adding one column; after the sum of the first column is obtained, add all except the right-hand figure of this sum to the second column, adding the second column the same way you added the first; proceed in like manner with all the columns, always adding to each successive column the sum of the column in the next lower order, minus the right-hand figure.*

N. B. The small figures which we place to the right of the column when adding are called *intergers*

The addition by intergers, or by forming a new column, as explained in the preceding examples, should be used only in adding very long columns of figures, say a long ledger column, where the footings of each column would be two or three hundred, in which case it is superior and much more easy than any other mode of addition; but in adding short columns it would be useless to form an extra column, where there is only, say six or eight figures to be added. In making short additions, the following suggestions will, we trust, be of use to the accountant who seeks for information on this subject.

In the addition of several columns of figures, where there are only four or five deep, or when their respective sums will range from twenty-five to forty, the accountant should commence with the unit column, adding the sum of the first two figures to the sum of the next two, and so on, naming only the results, that is, the sum of every two figures.

In the present example, in adding the unit column instead of saying 8 and 4 are 12 and 5 are 17 and 6 are 23, it is better to let the eye glide up the column, reading only, 8, 12, 17, 23; and still better, instead of making a separate addition for each figure, group the figures thus: 12 and 11 are 23, and proceed in like manner with each column. For short columns this is a very expeditious way, and indeed to be preferred, but for long columns, the addition by integers is the most useful, as the mind is relieved at intervals, and the mental labor of retaining the whole amount, as you add, is avoided, which is very important to any person whose mind is constantly employed in various commercial calculations

In adding a long column, where the figures are of a medium size, that is, as many 8s and 9s as there are 2s and 3s, it is better to add about three figures at a time, because the eye will distinctly see that many at once, and the ingenious student will in a short time, if he adds by integers, be able to read the amount of three figures at a glance, or as quick, we might say, as he would read a single figure.

Here we begin to add at the bottom of the unit column and add successively three figures at a time, and place their respective sums, minus 10, to the right of the last figure added; if the three figures do not make 10, add on more figures; if the three figures make 20 or more, only add two of the figures. The little figures that are placed to the right and left of the column are called integers. The integers in the present example, belonging to the units' column, are 4, 4, 5, — 4, 6, which we add together making 23; place down 3 and add 2 to the number of integers, which gives 7, which we add to the tens and proceed as before.

$$
\begin{array}{r}
^{5}26^{6} \\
67 \\
43 \\
38^{4} \\
^{6}54 \\
62 \\
87^{6} \\
^{4}65 \\
53 \\
44^{4} \\
^{8}77 \\
33 \\
84^{4} \\
^{8}56 \\
14 \\
\hline
803
\end{array}
$$

REASON.—In the above example, every time we placed down an integer we discarded a ten, and when we set down the 3 in the answer we discarded two tens; hence, we add 2 on to the number of integers to ascertain how many tens were discarded; there being 5 integers, it made 7 tens, which we now add to the column of tens; on the same principle we might add between 20 and 30, always setting down a figure before we got to 30; then every integer set down would count for 2 tens, being discarded in the same way, it does in the present instance for one ten. When we add between 10 and 20, and in very long columns, it would be much better to go as near 30 as possible, and count 2 tens for every integer set down, in which case we would set down about one-half as many integers as when we write an integer for every ten we discard

When adding long columns in a ledger or day-book, and where the accountant wishes to avoid the writing of extra figures in the book, he can place a strip of paper alongside of the column he wishes to add, and write the integers on the paper, and in this way the column can be added as conveniently almost as if the integers were written in the book.

Perhaps, too, this would be as proper a time as any other to urge the importance of another good habit; I mean *that of making plain figures.* Some persons accustom themselves to making mere scrawls, and important blunders are often the result. If letters be badly made, you may judge from such as are known; but if one figure be illegible, its value cannot be inferred from the others. The vexation of the man who wrote for 2 or 3 monkeys, and had 203 sent him, was of far less importance than errors and disappointments sometimes resulting from this inexcusable practice.

We will now proceed to give some methods of proof. Many persons are fond of proving the correctness of work, and pupils are often instructed to do so, for the double purpose of giving them exercise in calculation and saving their teacher the trouble of reviewing their work.

There are special modes of proof of elementary operations, as by casting out threes or nines, or by changing the order of the operation, as in adding upward and then downward. In addition, some prefer reviewing the work by performing the Addition downward, rather than repeating the ordinary operation. This is better, for if a mistake be inadvertently made in any calculation, and the same routine be again followed, we are very liable to fall again into the same error. If, for instance, in running up a column of Addition you should say 84 and 8 are 93, you would be liable, in going over the same again, in the same way to slide insensibly into a similar error; but by beginning at a different point this is avoided.

This fact is one of the strongest objections to the plan of cutting off the upper line and adding it to the sum of the rest, and hence some cut off the lower line by which the spell is broken. The most thoughtless cannot fail to see that adding a line *to* the sum of the rest is the same as adding it in *with* the rest.

The mode of proof by casting out the nines and threes will be fully explained in a following chapter.

A very excellent mode of avoiding error in adding long columns is to set down the result of each column on some waste spot, observing to place the numbers successively a place further to the left each time, as in putting down the product figures in multiplication; and afterward add up the amount. In this way if the operator lose his count, he is not compelled to go back to units, but only to the foot of the column on which he is operating. It is also true that the brisk accountant, who thinks on what he is doing, is less liable to err than the dilatory one, who allows his mind to wander. Practice, too, will enable a person to read accounts without naming each figure: thus, instead of saying 8 and 6 are 14, and 7 are 21 and 5 are 26, it is better to let the eye glide up the column, reading only 8, 14, 21, 26, etc.; and, still further, it is quite practicable to accustom one's self to group the figures in adding, and thus proceed very rapidly. Thus in adding the units' column, 62 instead of adding a figure at a time, we see at a glance that 4 — and 2 are 6, and that 5 and 3 are 8; then 6 and 8 are 14; we may then, if expert, add constantly the sum of two or three figures at a time, and with practice this will be found highly advantageous in long columns of figures; or two or three columns may be added at a time, as the practised eye will see that 24 and 62 are 86 almost as readily as that 4 and 2 are 6.

87
23
45
62
24

MULTIPLICATION.

Multiplication, in its most general sense, is a series of additions of the same number; therefore, in multiplication, a number is repeated a certain number of times, and the result thus obtained is called the product. When the multiplicand and the multiplier are each composed of only two figures, to ascertain the product, we have the following

Rule. — *Set down the smaller factor under the larger, units under units, tens under tens. Begin with the unit figure of the multiplier, multiply by it, first the units of the multiplicand, setting the units of the product, and reserving the tens to be added to the next product; now multiply the tens of the multiplicand by the unit figure of the multiplier, and the units of the multiplicand by tens*

figure of the multiplier; add these two products together, setting down the units of their sum, and reserving the tens to be added to the next product; now multiply the tens of the multiplicand by the tens' figure of the multiplier, and set down the whole amount. This will be the complete product.

Remark.—Always add in the tens that are reserved as soon as you form the first product.

EXAMPLE 1.—EXPLANATION.

1. Multiply the units of the multiplicand by the unit figure of the multiplier, thus: 1 ×4 is 4; set the 4 down as in example. 2. Multiply the tens in the multiplicand by the unit figure in the multiplier, and the units in the multiplicand by the tens figure in the multiplier, thus: 1×2 is 2; 3×4 are 12, add these two products together, 2 plus 12 are 14, set the 4 down as in example, and reserve the 1 to be added to the next product. 3. Multiply the tens in the multiplicand by the tens' figures in the multiplier, and add in the tens that were reserved, thus; 3×2 are 6, and 6 plus 1 equal 7; now set down the whole amount, which is 7.

24
31
—
744

EXAMPLE 1.—EXPLANATION.

Multiply first upper by units, 5×3 are 15, set down the 5, reserve the 1 to carry to the next product; now multiply second upper by units and first upper by tens, 5×2 are 10, plus 1 are 11, 4×3 are 12, add these products together; 11 plus 12 are 23, set down the 3, reserve the 2 to carry; now multiply third upper by units, and second upper by tens, add these two products together, always adding on the re-

123
45
—
5535

served figure to the first product; 5 ×1 are 5, plus 2 are 7, 4×2 are 8, and 7 plus 8 are 15, set down the 5, reserve the 1; now multiply third upper by tens, and set down the whole amount; 4×1 are 4 plus 1 are 5, set down the 5. This will give the complete product.

Multiply 32 by 45 in a single line. Here we multiply 5×2 and set down and carry as usual; then to what you carry add 5×3 and 4× 2, which gives 24; set down 4 and carry 2 to 4×3, which gives 14 and completes the product.

           32
           45
      ————
    1440

Multiply 123 by 456 in a single line.

Here the first and second places are found as before; for the third, add 6×1, 5×2, 4× 3, with the 2 you had to carry, making 30; set down 0 and carry 3; then drop the units' place and multiply the hundreds and tens crosswise, as you did the tens and units, and you find the thousand figure; then, dropping both units and tens, multiply the 4×1, adding the 1 you carried, and you have 5, which completes the product. The same principle may be extended to any number of places; but let each step be made perfectly familiar before advancing to another. Begin with two places, then take three, then four, but always practising some time on each number, for any hesitation as you progress will confuse you.

     123
     456
  ————
 56088

CURIOUS AND USEFUL CONTRACTIONS.

To multiply any number, of two figures, by 11.

RULE.—*Write the sum of the figures between them.*

1. Multiply 45 by 11. Ans. 495.

Here 4 and 5 are 9, which write between 4 and 5.

2. Multiply 34 by 11. Ans. 374.

N. B. When the sum of the two figures is over 9, increase the left-hand figure by the 1 to carry.

3. Multiply 87 by 11. Ans. 957.

To square any number of 9s instantaneously, and without multiplying.

RULE.—*Write down as many 9s less one as there are 9s in the given number, an 8, as many 0s as 9s, and a 1.*

4. What is the square of 9999? Ans. 99980001.

EXPLANATION.—We have four 9s in the given number, so we write down three 9s, then an 8, then three 0s, and a 1.

5. Square 999999. Answer 999998-000001.

To square any number ending in 5.

RULE.—*Omit the 5 and multiply the number as it will then stand by the next higher number, and annex 25 to the product.*

6. What is the square of 75? Ans. 5625.

EXPLANATION. — We simply say, 7 times 8 are 56, to which we annex 25.

7. What is the square of 95? Ans. 9025.

PRACTICAL BUSINESS METHOD

For Multiplying all Mixed Numbers.

Merchants, grocers, and business men generally, in multiplying the mixed numbers that arise in the practical calculations of their business, only care about having the answer correct to the nearest cent; that is, they disregard the fraction. When it is a half cent or more, they call it another cent, if less than half a cent, they drop it. And the object of the following rule is to show the business man the easiest and most rapid process of finding the product to the nearest unit of any two numbers, one or both of which involves a fraction.

GENERAL RULE.

To multiply any two numbers to the nearest unit.

1st. *Multiply the whole number in the multiplicand by the fraction in the multiplier to the nearest unit.*

2d. *Multiply the whole nmmber in the multiplier by the fraction in the multiplicand to the nearest unit.*

3d. *Multiply the whole numbers together and add the three products in your mind as you proceed.*

N. B. In actual business the work can generally be done mentally, for only easy fractions *occur in business.*

N. B. This rule is so simple and so true, according to all business usage, that every accountant should make himself perfectly familiar with its application. There being no such thing as a fraction to add in, there is scarcely any liability to error or mistake. By no other arithmetical process can the result be obtained by so few figures.

EXAMPLE FOR MENTAL OPERATION.

Multiply 11⅓ by 8¼ by business method.

Here ¼ of 11 to the nearest unit is 3, and ⅓ of 8 to the nearest unit is 3, making 6, so we simply say, 8 times 11 are —— 88 and 6 are 94. Ans.   94

11⅓
8¼

REASON.—¼ of 11 is nearer 3 than 2, and ⅓ of 8 is nearer 3 than 2, Make the nearest whole number the quotient.

Retail merchants, in buying goods by wholesale, buy a great many articles by the dozen, such as boots and shoes, hats and caps, and notions of various kinds. Now, the merchant, in buying, for instance, a dozen hats, knows exactly what one of those hats will retail for in the market where he deals; and, unless he is a good accountant, it will often take him some time to determine whether he can afford to purchase the dozen hats and make a living profit in selling them by the single hat; and in buying his goods by auction, as the merchant often does, he has not time to make the calculation before the goods are cried off. He, therefore, loses the chance of making good bargains by being afraid to bid at random, or if he bids, and the goods are cried off, he may have made a poor bargain by bidding thus at a venture. It then becomes a useful and practical problem to determine *instantly* what per cent. he would gain if he retailed the hats at a certain price.

RAPID PROCESS OF MARKING GOODS

To tell what an article should retail for to make a profit of 20 per cent. is done by removing the decimal point one place to the left.

For instance, if hats costs $17.50 per dozen, remove the decimal point one place to the left, making $1.75, what they should be sold for a piece to gain 20 per cent. on the cost. If they cost $31.00 per dozen, they should be sold for $3.10 apiece, etc.

We take 20 per cent as the basis, for the following reasons, namely: because we can determine instantly, by simply removing the decimal point, without changing a figure; and, if the goods would not bring at least 20 per cent. profit in the home market, the merchant could not afford to purchase and would look for goods at lower figures.

Now, as removing the decimal point one place to the left, on the cost of a dozen articles gives the selling price of a single one with 20 per cent. added to the cost, and, as the cost of any article is 100 per cent., it is obvious that the selling price would be 20 per cent. more, or 120 per cent; hence, to find 50 per cent. profit, which would make the selling price 150 per cent., we would first find 120 per cent., then add 30 per cent., by increasing it one-fourth itself; to make 40 per cent., add 20 per cent., by increasing it one-sixth itself; for 35 per cent. increase it one-eighth itself, etc. Hence, to mark an article at any per-cent. profit, we have the following

GENERAL RULE.

First find 20 *per cent. profit by removing the decimal point one place to the left on the price the articles cost a dozen; then, as* 20 *per cent. profit is* 120 *per cent., add to, or substract from, this amount the fractional part that the required per cent. added to* 100 *is more or less than* 120.

TABLE.

For Marking all Articles bought by the Dozen.

N. B. Most of these are used in business.
To make 20 pr ct. remove the point one place to the left.

"	80	"	"	" and add one-half itself.
"	60	"	"	" one-third "
"	50	"	"	" one-fourth "
"	44	"	"	" one-fifth "
"	40	"	"	" one-sixth "
"	37½	"	"	" one-seventh "
"	35	"	"	" one-eighth "

To make 33⅓ p ct. remove point and add one-ninth itself.				
"	32	"	"	" " one-tenth "
"	30	"	"	" one-twelfth "
"	28	"	"	" one-fifteenth "
"	26	"	"	" one-twentieth '
"	25	"	"	"one-twent⁴- fourth
"	12½	"	"	subtract one-sixteenth "
"	16⅔	'	"	" one-thirty-sixth
"	18¾	"	"	" one-ninety-sixth

If I buy 1 doz. shirts for $28.00, what shall I retail them for to make 50 per ct.? Ans. $3.50.

EXPLANATION.—Remove the point one place to the left, and add on ¼ itself.

Where the Multiplier is an Aliquot part of 100.

Merchants in selling goods generally make the price of an article some aliquot part of 100, as in selling sugar at 12½ cents a pound or 8 pounds for 1 dollar, or in selling calico, for 16 2-3 cents a yard or 6 yards for 1 dollar, etc. And to become familiar with all the aliquot parts of 100, so that you can apply them readily when occasion requires, is perhaps the most useful, and, at the same time, one of the easiest arrived at of all the computations the accountant must perform in the practical calculations of the counting-room.

TABLE.

Of the Aliquot parts of 100 *and* 1000.

N. B. Most of these are used in business.

12½ is ⅛ part of 100.	8⅓ is 1-12 part of 100
25 is 2-8 or ¼ of 100.	16⅔ is 2-12 or 1-6 of 100
37½ is 3-8 part of 100.	33⅓ is 4-12 or ⅓ of 100
50 is 4-8 or ½ of 100.	66⅔ is 8-12 or ⅔ of 100
62½ is ⅝ part of 100.	88⅓ is 10-12 or 5-6 of 100
75 is 6-8 or ¾ of 100.	125 is ⅛ part of 1000
87½ is ⅞ part of 100.	250 is 2-8 or ¼ of 1000
6¼ is 1-16 part of 100.	375 is ⅜ part of 1000
18¾ is 3-16 part of 100.	625 is ⅝ part of 1000
31¼ is 5-16 part of 100.	875 is ⅞ part of 1000

To multiply by an aliquot part of 100.

RULE.—*Add two ciphers to the multiplicand, then take such part of it as the multipliers is part of* 100.

N. B. If the multiplicand is a mixed number reduce the fraction to

a decimal of two places before dividing.

General Rules for Cancellation.

RULE 1ST. Draw a perpendicular line ; observe this line represents the sign of equality. On the right-hand side of this line place dividends only ; on the left hand side place divisors only ; having placed dividends on the right and divisors on the left as above directed,

2d. Notice whether there are ciphers both on the right and left of the line ; if so, erase an equal number from each side.

3d. Notice whether the same number stands both on the right and left of the line ; if so, erase them both.

4th. Notice again if any number on either side of the line will divide any number on the opposite side without a remainder ; if so, divide and erase the two numbers, retaining the quotient figure on the side of the larger number.

5th. See if any two numbers, one on each side, can be divided by any assumed number without a remainder ; if so, divide them by that number, and retain only their quotients. Proceed in the same manner as far as practicable, then,

6th. Multiply all the numbers remaining on the right-hand side of the line for a dividend, and those remaining on the left for a divisor.

7th. Divide, and the quotient is the answer.

SIMPLE INTEREST BY CANCELLATION.

RULE. — *Place the principal, time and rate per cent. on the right-hand side of the line. If the time consists of years and months, reduce them to months, and place* 12 *(the number of months in a year) on the left-hand side of the line. Should the time consist of months and days, reduce them to days or decimal parts of a month. If reduced to days, place* 36 *on the left. If to decimal parts of a month, place* 12 *only, as before.*

Point off two decimal places when the time is in months, and three decimal places when the time is in days.

NOTE.—If the principal contains cents, point off four decimal places when the time is in months, and five decimal places when the time is in days.

NOTE. — *We place* 36 *on the left because there are* 360 *interest days in a year. (Custom has made this lawful.)*

LIGHTNING METHOD OF COMPUTING INTEREST.

On all notes that bear $12 *per annum, or any aliquot part or multiple of* $12.

If a note bears $12 per annum, it will certainly bear $1 per month : hence the time in months would be the interest in $; and the decimal parts of a month would be the interest in decimal parts of a $; therefore when the note bears $12 per annum we have the following rule:

RULE.—*Reduce the years to months, add in the given months, and place one-third of the days to the right of this number, and you have the interest in dimes.*

EXAMPLE 1.—Required the interest of $200 for 3 years, 7 months, and 12 days, at 6 per cent.

$$200$$
$$6$$

⅓ of 12 days = 4.

$$\text{Yr. Mo. Da.}$$
$12.00 = \text{int. for 1 yr.} \quad 3 \quad 7 \quad 12 = 43.\ 4\ \text{mo.}$
Hence 43.4 dimes. or $43.40cts., *Ans*

We see by inspection that this note bears $12 interest a year; hence the time reduced to months, with one-third of the days to the right, is the interest in dimes. If this note bore $6 a year, instead of $12, we would take one-half of the above interest; if it bore $18 instead of $12, we would add one-half; if it bore $24, instead of $12, we would multiply by 2, etc.

EXAMPLE 2. — Required the interest of $150 for two years, 5 months, and 13 days, at 8 per cent.

150
8
———
⅓ of 13 days =4⅓
Yr. Mo. Da.
$12.00 = int. for 1 yr. 2 5 13 = 29. 4⅓mos.

Hence $29. 4⅓ dimes, or $29.433⅓cts., *Ans.*

We see by inspection that this note bears $12 interest a year; hence the time reduced to months, with one-third of the days placed to the right, gives the interest at once.

EXAMPLE 3.—Required the interest of $160 for 11 years, 11 months, and 11 days, at 7½ per cent.

160
7½
———
⅓ of 11 days = 3⅔·
Yr. Mo. Da.
$12.00 = int. for 1 yr. 11 11 11 = 143⅔mos.

Hence 143.3⅔ dimes, or $143.36⅔cts., *Ans.*

When the interest is more or less than $12 a Year.

RULE.—*First find the interest for the given time on the base of $12 interest a year; then, if the interest on the note is only $6 a year, divide by 2; if $24 a year, multiply by 2; if $18 a year, add on one-half, etc.*

EXAMPLE 1.—What is the interest of $300 for 4 years, 7 months, and 18 days, at 6 per cent.?

⅓ of 18 days =6.
4yr. 7mo. 18da.=55.6mo.
300
6
———
$18.00 =int. for 1 yr. 2)55.6, int. at $12 a yr.
$18=1½ times $12.   27.8.

$83.4. *Ans.*

If the interest was $12 a year, $55.60 would be the answer; because 55.6 is the time reduced to months; but it bears $18 a year, or 1½ times 12; hence 1½ times 55.6 gives the interest at once.

EXAMPLE 2.—Required the interest of $150 for three years, 9 months, and 27 days, at four per cent.

⅓ of 27 days =9.
150   3yr. 9mo. 27da = 45.9mo.
4   2)45.9. int. at $12 a year.
———
$6.00 = int. for 1 yr. $22.95, *Ans.*
$6 = ⅓ times $12.

If the interest was $12 a year, $45.90 would be the answer; because $45.9 is the time reduced to months ; but it bears $6 a year, or ½ times 12; hence ½ times 45.9 gives the interest at once.

———

RULES FOR DETERMINING THE WEIGHT OF LIVE CATTLE.

Measure in inches the girth round the breast, just behind the shoulder-blade, and the length of the back from the tail to the fore part of the shoulder-blade. Multiply the girth by the length, and divide by 144. If the girth is less than three feet, multiply the quotient by 11; if between three feet and five feet, multiply by 16; if between five feet and seven feet, multiply by 23; if between seven and nine feet, multiply by 31. If the animal is lean, deduct 1-20th from the result.

Take the girth and length in feet, multiply the square of the girth by the length, and multiply the product by 3.36. The result will be the answer in pounds. The live weight, multiplied by 605, gives a near approximation to the net weight.

ASTRONOMICAL CALCULATIONS.

A scientific method of telling immediately what day of the week any date transpired or will transpire, from the commencement of the Christian Era, for the term of three thousand years.

MONTHLY TABLE.

The ratio to add for each month will be found in the following table:

Ratio of June is.........0	Ratio of October is.......3
Ratio of September is....1	Ratio of May is..........4
Ratio of December is.....1	Ratio of August is.......5
Ratio of April is.........2	Ratio of March is........6
Ratio of July is..........2	Ratio of February is....6
Ratio of January is......3	Ratio of November is....6

NOTE. — On Leap Year the ratio of January is 2, and the ratio of February is 5. The ratio of the other ten months do not change on Leap Years.

CENTENNIAL TABLE.

The ratio to add for each century will be found in the following table:

Christian Era.						
200,	900,	1800,	2200,	2600,	3000, ratio is......0
300,	1000,	....	....	....	.... ratio is......6
400,	1100,	1900,	2300,	2700,	.... ratio is......5
500,	1200,	1600,	2000,	2400,	2800, ratio is......4
600,	1300,	....	....	....	.... ratio is......3
000,	700,	1400,	1700,	2100,	2500, 2900 ratio is......2
100,	800,	1500,	....	....	.... ratio is......1

NOTE. — The figure opposite each century is its ratio; thus the ratio for 200, 900, etc., is 0. To find the ratio of any century, first find the century in the above table, then run the eye along the line until you arrive at the end, the small figure at the end is its ratio.

METHOD OF OPERATION.

RULE.* — *To the given year add its fourth part, rejecting the fractions;*

*When dividing the year by 4, always leave off the centuries. We divide by 4 to find the number of Leap Years.

to this sum add the day of the month; then add the ratio of the month and the ratio of the century. Divide this sum by 7; the remainder is the day of the week counting Sunday as the first, Monday as the second, Tuesday as the third, Wednesday as the fourth, Thursday as the fifth, Friday as the Sixth, Saturday as the seventh; the remainder for Saturday will be O or zero.

EXAMPLE 1.—Required the day of the week for the 4th of July, 1810.

To the given year, which is........................10	
Add its fourth part, rejecting fractions............ 2	
Now add the day of the month, which is............ 4	
Now add the ratio of July, which is................ 2	
Now add the ratio of 1800, which is............... 0	
Divide the whole sum by 7	7	18-4
	2

We have 4 for a remainder, which signifies the fourth day of the week, or Wednesday.

Rule for finding the number of feet of boards which can be cut from any log whatever.

From the diameter of the log, in inches, substract 4 for the slabs and saw-calf. Then multiply the remainder by half itself and the product by the length of the log in feet, and divide the result by 8; the quotient will be the number of square feet.

EXAMPLE 1.—What is the number of feet of boards which can be cut from a log 24 inches in diameter and 12 feet long?

Diameter........................24 inches	
For slabs and saw-calf...........4	
Remainder......................20	
Half remainder.................10	
	200
Length of log....................12	
	8	2400
	300 the number of feet.

HENDERSON'S LIGHTNING PROCESS,

*For Computing Time and Interest, Squaring and Multiplying Numbers, and a
Fine Method for Dividing Fractions, and an infinite number of
of ways of Extracting Square and Cube Root.*

The following Table gives the Interest on any amount at 7 per cent., by
simply removing the point to right or left, as the case may require:

Number of Days.	$100	$90	$80	$70	$60	$50	$40	$30	$20
1....	.0192	.01726	.01534	.01342	.01151	.00950	.00767	.00575	.00384
2....	.0384	.03452	.03058	.02685	.02301	.01918	.01534	.01151	.00767
3....	.0575	.05178	.04603	.04027	.03452	.02877	.02301	.01726	.01151
4....	.0767	.06904	.06137	.05370	.04603	.02836	.03068	.02301	.01536
5....	.0959	.08630	.07671	.06712	.05753	.04795	.03836	.02877	.01918
6....	.1151	.10356	.09205	.08055	.06904	.05753	.04603	.03452	.02313
7....	.1342	.12082	.10740	.09897	.08055	.06712	.05370	.04027	.02685
8....	.1532	.13808	.12274	.10740	.09205	.07671	.06137	.04603	.03068
9....	.1726	.15534	.13808	.12089	1.0356	.08630	.06904	.05178	.03452
90....	1.7260	1.5342	1.38082	1.20822	1.03562	.86301	.69041	.51781	.34521
93....	1.7836	1.60521	1.42685	1.24849	1.07014	.89178	.71342	.53508	.35671
100....	1.9178	1.82603	1.53425	1.24247	1.15065	.95890	.76712	.57534	.48356

For 10-7 of a year remove the decimal point one place to the left; 1-7, or
52 days, two places to the left. Increase or diminish the results to suit the
time..

When the Rate is 6 per cent.

For 5-3 of a year, or 20 months, remove the point one place to the left; 60
days, two places, and 6 days three places to the left.

$5.|00	$ 94|7.50
7.|50.25	345|8.50
8.|36.50	94|3.20
9.|47.75 Is the interest at 7 per cent. for 52 days, or 1-7 of a year.	64|9.50 At 6 per cent. for 20 months; for 60 days draw the line two places to the left of the decimal point; and for 6 days three places, etc.

When the rate is 5 per cent.

For two years remove the point one place to the left, and 73 days two places
to the left.

When the rate is 7½ per cent.

For 4-3 of a year or 16 months remove the point one place; for 48 days
two places, the result modifying to suit the time given.

When the rate is 8 *per cent.*

For 15 months, remove the point one place ; for 1-8 of a year, or 45 days, two places to the left.

To MAKE a rule for all rates, divide 100 by the rate and the quotient is the time, when the principal equals the interest and the point remains the same ; divide 10 by the rate, and the quotient indicates the time or base you work from, when you remove the point 1 place to the left ; divide unity by the rate, and the result is the part of a year and the number of days, when the point is to be removed two places to the left.

To FIND THE INTEREST by the table, for any given time and any number of dollars, look on the Time Table for the time, and on the Interest Table for the interest of twenty, thirty and forty dollars, etc. Modify by removing the point right or left to suit the example given.

You can find the interest very conveniently by taking the number of months and $\frac{1}{3}$ of the days, and multiply that by $\frac{1}{2}$ of the principal, and you have the interest at 6 per cent. in cents.

RULE.—Remove the point one place to the left, because one-tenth of the principal equals the interest.

Remove the point two places, for one-hundredth of the principal equals the interest.

Remove the point three places, because one-thousandth of the principal equals the interest.

These methods give the interest of all finite sums of money, for the time and rate mentioned in each rule.

To reach all other time, increase or diminish the results to suit the time given.

Thus: $500 for 1-7 of a year at 7 per cent. is five dollars ; for one half of that time $2.50 ; for one fourth, $1.25, &c.; for one year it is seven times $5.00, $35.

$400, for 1-6 of a year, and rate 6, is $4 ; for one-half of that time, $2.

For one year $24 ; for 1-60 of a year, or 6 days, remove the point three places and the interest is 40 cents ; for one half of that time it is 20 cents.

The rule may thus be expressed : The reciprocal of the rate is the time when the point can be removed two places to the left in all cases ; ten times that time remove it one place to the left, one tenth of the same time three places to the left : Increase or diminish the results to suit the time given.

————

TO MULTIPLY NUMBERS, FIRST KNOW HOW TO SQUARE THEM.

(99)=9801 Take the complement of 99 from it, call it hundreds, and add the square of the complement:

$(101)2=10201$
$(102)2=10404$
$(103)2=10609$
&c., &c.

When above the base, add the supplement, call it hundreds, and increase it by the square of the supplement.

$(11)2=121$
$(12)2=144$
$(13)2=169$
&c., &c.

Then $n=99$
and $c=1$
$n+c=100$
$n-c=98$
$n2-c2=9800$

Now add c^3 to both members of the equation, and we have the square of the number. In the same manner, let n equal the number and s the supplement, and the reason of the rule becomes evident. For same reason:

$(98)2=9604$
$(97)2=9409$
$(96)2=9216$
$(95)2=9025$
&c., &c.

Take any number that is easy to multiply by for the base 10, 20, 40, 50, &c.

The product of any two numbers is the Square of the Mean diminished by the Square of Half the Difference.

$39\times41=(40)2—12=1599$
$38\times42=(40)2—22=1596$
$37\times43=(40)2—32=1591$
&c., &c.

From the square of the mean subtract the right hand digit of the greater number; because it indicates half of the difference of the two numbers.

$79\times81=6399$
$78\times82=6396$
&c., &c.

$$\frac{8\frac14}{6\frac12} = \frac{33}{26}$$

Multiply both dividend and divisor by the least common multiple of the denominators of the fractional parts.

$23\times27=621$
$22\times28=616$
$24\times26=624$
&c., &c.

Increase 2 by 1, and multiply by the other tens digit, and annex the product of unit's digits. Add 1, because the sum of the units digits is $=10$.

$8\ 1\text{-}2\times8\ 1\text{-}2=72\ 1\text{-}4$
$8\ 1\text{-}3\times8\ 2\text{-}3=72\ 2\text{-}9$
$8\ 2\text{-}5\times8\ 3\text{-}5=72\ 6\text{-}25$ &c., for all similar examples.

HENDERSON'S METHOD OF EXTRACTING CUBE ROOT.

```
10000      1953125(100+20+5
 ———       1000000
30000      ———
 6000       953125
  400       728000
 ———       ———
36400       225125
 6000       225125
  800      ———
 ———
43200
 1800
   25
 ———
45025
```

Add to each true divisor, as they occur, twice the surface of one side of the small cube, and one of each of the three parallelopipedons, for a trial divisor; because that will make three sides of the complete cube.

By observation the reason is evident and the conclusion just, for making trial and true divisors by this method.

We have an infinite number of ways of finding the square root, cube root, &c. Presume the root to be divided into a certain number of parts. Square the parts in square root; cube them in cube root to find the divisor. Thus let $a+a$ represent the square root of any number. The square of $a+a$ is 4 a^2: hence divide any number by 4 and extract the square root of the quotient, and we have half of the root. Divide any number by the square of 3, and extract the square root of the quotient, and we have one-third of the root, &c., for all numbers. In the cube root we cube the number representing the parts the root is divided into, for a divisor

To find the Day of the Week from the Day of the Month.

Cast the sevens out of the day of the month, the ratio of the month, the ratio of the year, and the year. One of a remainder will be the first day of the week; two second, &c., 0 the last day of the week. The ratio of the month is found above its name. The ratio of every month except January and February is one more in Leap Years.

Jan'y 3		Feb'y 6		March 6		April 2		May 4		June 0		July 2		August 5		Sept. 1		October 3		Novem. 6		Decem. 1	
1	1	1	32	1	60	1	91	1	121	1	152	1	182	1	213	1	244	1	274	1	305	1	335
2	2	2	33	2	61	2	92	2	122	2	153	2	183	2	214	2	245	2	275	2	306	2	336
3	3	3	34	3	62	3	93	3	123	3	154	3	184	3	215	3	246	3	276	3	307	3	337
4	4	4	35	4	63	4	94	4	124	4	155	4	185	4	216	4	247	4	277	4	308	4	338
5	5	5	36	5	64	5	95	5	125	5	156	5	186	5	217	5	248	5	278	5	309	5	339
6	6	6	37	6	65	6	96	6	126	6	157	6	187	6	218	6	249	6	279	6	310	6	340
7	7	7	38	7	66	7	97	7	127	7	158	7	188	7	219	7	250	7	280	7	311	7	341
8	8	8	39	8	67	8	98	8	128	8	159	8	189	8	220	8	251	8	281	8	312	8	342
9	9	9	40	9	68	9	99	9	129	9	160	9	190	9	221	9	252	9	282	9	313	9	343
10	10	10	41	10	69	10	100	10	130	10	161	10	191	10	222	10	253	10	283	10	314	10	344
11	11	11	42	11	70	11	101	11	131	11	162	11	192	11	223	11	254	11	284	11	315	11	345
12	12	12	43	12	71	12	102	12	132	12	163	12	193	12	224	12	255	12	285	12	316	12	346
13	13	13	44	13	72	13	103	13	133	13	164	13	194	13	225	13	256	13	286	13	317	13	347
14	14	14	45	14	73	14	104	14	134	14	165	14	195	14	226	14	257	14	287	14	318	14	348
15	15	15	46	15	74	15	105	15	135	15	166	15	196	15	227	15	258	15	288	15	319	15	349
16	16	16	47	16	75	16	106	16	136	16	167	16	197	16	228	16	259	16	289	16	320	16	350
17	17	17	48	17	76	17	107	17	137	17	168	17	198	17	229	17	260	17	290	17	321	17	351
18	18	18	49	18	77	18	108	18	138	18	169	18	199	18	230	18	261	18	291	18	322	18	352
19	19	19	50	19	78	19	109	19	139	19	170	19	200	19	231	19	262	19	292	19	323	19	353
20	20	20	51	20	79	20	110	20	140	20	171	20	201	20	232	20	263	20	293	20	324	20	354
21	21	21	52	21	80	21	111	21	141	21	172	21	202	21	233	21	264	21	294	21	325	21	355
22	22	22	53	22	81	22	112	22	142	22	173	22	203	22	234	22	265	22	295	22	326	22	356
23	23	23	54	23	82	23	113	23	143	23	174	23	204	23	235	23	266	23	296	23	327	23	357
24	24	24	55	24	83	24	114	24	144	24	175	24	205	24	236	24	267	24	297	24	328	24	358
25	25	25	56	25	84	25	115	25	145	25	176	25	206	25	237	25	268	25	298	25	329	25	359
26	26	26	57	26	85	26	116	26	146	26	177	26	207	26	238	26	269	26	299	26	330	26	360
27	27	27	58	27	86	27	117	27	147	27	178	27	208	27	239	27	270	27	300	27	331	27	361
28	28	28	59	28	87	28	118	28	148	28	179	28	209	28	240	28	271	28	301	28	332	28	362
29	29			29	88	29	119	29	149	29	180	29	210	29	241	29	272	29	302	29	333	29	363
30	30			30	89	30	120	30	150	30	181	30	211	30	242	30	273	30	303	30	334	30	364
31	31			31	90			31	151			31	212	31	243			31	304			31	365

SUGGESTIONS ON TEACHING ARITHMETIC.

Qualifications.—The chief qualifications requisite in teaching Arithmetic, as well as other branches, are the following:—A thorough knowledge of the subject; a love for the employment; and an aptitude to teach. These are indispensable to success.

Classification—Arithmetic, as well as other studies, should be taught in classes. This method saves much time, and thereby enables the teacher to devote more time to oral illustrations.

The action of mind upon mind is a potential stimulant to exertion, and cannot fail to create a zeal for the study. The mode of analyzing and reasoning of one scholar often suggests new ideas to others in the class.

In classification, those should be put together who possess as nearly equal capacities as possible. If any of the class learn faster than the others, they should be allowed to take extra study, or be furnished with additional examples to solve, so that the whole class may advance together.

The *Blackboard* should be one of the indispensables of the school-room. Not a recitation should pass without its use. When a principle is to be demonstrated or an operation explained, if done upon the blackboard, all can see and understand at once.

Recitation.—The first object in conducting a recitation should be to secure the attention of the class. This is done chiefly by blending life and variety with the exercise. Students generally loathe dullness, while animation and variety are their delight. Every example should be carefully analyzed; the "why and wherefore" of every step in the solution should be required, till the learner becomes perfectly familiar with the process of reasoning.

Thoroughness.—This should be the motto of every teacher; without it, the great objects of study are radically defeated. In securing this object, much advantage is derived from frequent reviews.

ARITHMETIC.

Arithmetic is the science of numbers, and the art of computing by them.

A *number* is a unit or a collection of units.

A *unit* is a single thing, or one.

Quantity is anything that can be increased, diminished, or measured.

The *fundamental* rules of Arithmetic are *Addition, Subtraction, Multiplication,* and *Division*.

NUMERATION.

Numeration is the process of reading numbers when expressed by figures.

Figures are characters used to express numbers in arithmetic. There are ten: 1, 2, 3, 4, 5, 6, 7, 8, 9, 0. Each figure has two values—*simple* and *local*. The *simple* value is that expressed by the figure when standing alone or in the unit's place.

The *local* value is that expressed when connected with other figures, and depends upon its distance from the unit's place.

The *cipher* denotes the absence of something, and when placed to the right of a figure, it increases the value of that figure *ten times*, or multiplies it by *ten*.

The value of a figure is illustrated by the following numeration table:—

Trillions, &c.	Hundreds of Billions,	Tens of Billions,	Billions,	Hundreds of Millions,	Tens of Millions,	Millions,	Hundreds of Thousands,	Tens of Thousands,	Thousands,	Hundreds,	Tens,	Units,
4	3	2	1	9	8	7	6	5	4	3	2	1

This table may be run on to Quadrillions, Quintillions, Sextillions, Septillions, Octillions, Nonillions, Decillions, Undecillions, Duodecillions, Tredecillions, Quatuordecillions, Quindecillions, Sexdecillions, Sepdecillions, Ocdecillions, Nondecillions, Vigintillions, etc.

APPENDIX TO LIGHTNING CALCULATOR,

CONTAINING GENERAL INFORMATION.

ARITHMETICAL SIGNS.

= Sign of equality 5 = 5, read five equals five.

+ Plus, the sign of addition; 4+8 =12, read four plus eight equal twelve.

— Minus, the sign of subtraction; 9— 3 = 6, read nine minus three equals six.

× Sign of multiplication; 4 × 2= 8, read four multiplied by two equals eight.

÷ Sign of division; 12÷3=4, read twelve divided by three equal four.

√ Radical sign, or sign of square root, when placed over a number, signifies that the square root is to be extracted; √ 64=8, read the square root of sixty-four equals eight.

∛ Sign of cube root, shows that the cube root is to be extracted; ∛ 27=3, read the cube root of twenty-seven equals three.

SHORT METHODS OF MULTIPLICATION AND DIVISION

To multiply by 25.

1. Multiply 492 by 25.

Operation.
4|49200
———
Analysis.—By annex- |12300   Ans.
ing two ciphers to the multiplicand, we multiply it by 100, which gives a product four times too great, as the multiplier, 25, is but one-fourth of 100. Hence, annex two ciphers to the multiplicand, and divide by 4.

2. What will 45 acres of land cost at $25 an acre?

Ans. $1125.

To multiply by 12½.

Annex two ciphers and divide by 8.
1. Multiply 38 by 12½. *Operation*

8|3800
———
Analysis.—Annexing | two ciphers in this case | 475 Ans. produces a product eight times too great; hence divide by 8 to obtain one-eighth.

2. What will 13 yards calico cost at 12½ cents per yard?   Ans. $1.62½.

3. Multiply 12½ by 143. Ans. 1787½

To multiply by 33⅓

Annex two ciphers and divide by 3.
1. What will 27 yards cloth cost at 33⅓ cents per yard?   Ans. $9.
2. Multiply 33⅓ by 14.   Ans. 466⅔.
3. Multiply 47 by 33⅓.   Ans. 1566⅔.

To multiply by 125.

Annex three ciphers and divide by 8.
1. Multiply 568 by 125.

Operation.
8|568000
———
Analysis.—The stu- | dent will readily per- | 71000 Ans. ceive that by annexing three ciphers he multiplies by 1000; and that 125 is one-eighth of 1000. Hence divide by 8.

2. What will 44 pair of shoes cost at $1.25 per pair?   Ans. $55.00.
3. What will 125 yards cloth cost at 58 cents per yard?   Ans. $72.50.

To multiply any number ending in 5 by itself; that is, to square such a number.

1. Multiply 25 by 25.   *Operation.*

```
      25
      25
     ----
     625 Ans.
```

RULE.—*Square the 5 and set down the result, 25, then multiply the left-hand figure by the next figure above it in the order of numbers and prefix the product to the 25, and you have the correct result.*

NOTE.—This rule is based upon the well-known principle of squaring any number consisting of tens and units. The square of such a number is equal to the square of the tens, plus twice the product of the tens by the units, plus the square of the units.

Thus 25 is composed of 2 tens or 20 and 5 units; and the 20 squared equals,

```
                                         400
20 multiplied by 5 equals 100, twice
that is                                  200
And lastly, the 5 squared equals          25
                                        ----
                                         625
```

Again, 5 squared is 25, and the 2 multiplied by 3 (the next figure above it) is 6, which placed before the 25 makes 625.

2. Multiply 45 by 45.   Or thus:—

Operation.

```
                                    45
                                    45
5 squared equals 25, and           ---
five times 4 equals 20, and 20     225
prefixed to 25 makes 2025,         180
Ans.                              ----
                                  2025
```

3. Square 85.   Ans. 7225.

4. What will 65 yards of cloth cost at 65 cents per yard?   Ans. 42.25.

5. What shall I pay for 75 acres of land at $75 per acre?   Ans. $5625.

To multiply a mixed number ending in $\frac{1}{2}$ by itself.

RULE.—*Multiply the figure or number by the next figure above it, and place the square of $\frac{1}{2}$ (which is $\frac{1}{4}$ always) to the right.*

1. Multiply $2\frac{1}{2}$ by $2\frac{1}{2}$.   Ans. $6\frac{1}{4}$.
Operation.—The 2 multiplied by 3 is 6, and a half times a half is $\frac{1}{4}$; hence $6\frac{1}{4}$ is the product.

2. Multiply $4\frac{1}{2}$ by $4\frac{1}{2}$.   Ans. $20\frac{1}{4}$.
Say five times 4 is 20 and annex the $\frac{1}{4}$.

3. What will $12\frac{1}{2}$ yards of cloth cost at $12\frac{1}{2}$ cents per yard?   Ans. 1.56\frac{1}{4}$.
Say 13 times 12. So of any other numbers.

To square any mixed number that terminates with $\frac{1}{4}$.

RULE.—*Square the $\frac{1}{4}$; then square the whole number, add half the whole number is its square, and place the result to the left of the $\frac{1}{16}$.*

1. What is the square of $4\frac{1}{4}$? Ans. $18\frac{1}{16}$.
Operation. — $\frac{1}{4}$ squared is $\frac{1}{16}$; 4 squared is 16 and 2 added makes 18. Then $18\frac{1}{16}$.

2. What is the square of $6\frac{1}{4}$?
Ans. $39\frac{1}{16}$.

3. What is the square of $12\frac{1}{4}$?
Ans. $150\frac{1}{16}$.

4. What is the square of $8\frac{1}{4}$?
Ans. $68\frac{1}{16}$.

SHORT METHODS OF DIVISION.

To divide by 25.

1. Divide 425 by 25.   Ans. 17.
Operation.

```
425
  4
-----
17.00
```

or

```
25 | 425
 4 |   4
---|------
1.00 | 17.00
     -------
      17 Ans.
```

Analysis.—By multiplying both divisor and dividend by 4, the divisor becomes 100, which enables us to perform the division by simply cutting off two figures from the right.

RULE. — *Multiply both divisor and dividend by that number which will change the divisor to a number of tens, hundreds, or thousands, and then divide by simply cutting off figures from the right.*

2. Divide 3489 by 25.

Ans. 139.56 or $\frac{14}{25}$.

Simply multiply by 8 and point off two, as $12\frac{1}{2}$ is one eighth of 100.

3. Divide 4800 by $12\frac{1}{2}$ Ans. 384.

4. Divide 54500 by 125.   Ans. 436.

Multiply by 8 and cut of *three*.

5. Divide 5470 by 250.

Ans. 21.880 or $21\frac{22}{25}$.

Simply multiply by 4 and point off *three* figures, and to the right is thousandths.

6. Divide 2000 by $333\frac{1}{3}$ Ans. 6.

Multiply by 3 and cut off three, it being the *third* of a thousand. And so of other numbers.

7. Divide $1400 by $33\frac{1}{3}$.   Ans. 42.

8. Divide 155 by $33\frac{1}{3}$.   Ans. 4.65.

9. Divide 1500 by $33\frac{1}{3}$.   Ans. 45.

PRACTICAL PROBLEMS ANALYZED.

1. Divide $140 among three boys, B to have twice as much as A, and C twice as much as B.

Ans. A 20, B 40, and C 80.

Operation.

$1 \times 20 = 20$ A
$2 \times 20 = 40$ B
$4 \times 20 = 80$ C

$\frac{7|140}{\quad|\ 20}$

If A gets 1 dollar, B 2, and C. 4,— all together they will get 7 dollars; but 7 in 140 *twenty times*, therefore each will get 20 times the index number; hence multiply each by 20.

2. If J. W. Butler sells three pairs of shoes for $12.33; the second twice as much as the first, and the third twice as much as both the others; what does he get for each pair?

Ans. $1.37, $2.74, $8.22.

3. If I have $\frac{2}{3}$ of my money in one pocket, $\frac{4}{5}$ in a second, and eight dollars in a third, how much money have I?

Ans. $180.

Operation.— $\frac{2}{3} + \frac{5}{9} = \frac{43}{45}$; $1 - \frac{43}{45} = \frac{2}{45}$.

Then, 8 divided by $\frac{2}{45} = 180$, Ans.

In such examples we merely get the sum of the fractional parts, substract it from 1, and divide the given number by the fractional remainder.

4. E. P. Williams has his goats in five fields, in the first he had $\frac{1}{4}$ of them, in the second $\frac{1}{3}$, in the third $\frac{1}{6}$, in the fourth $\frac{1}{12}$, and in the fifth 225, how many had he?   Ans. 600.

5. John Butler leaves his son an estate, $\frac{1}{3}$ of which he spends in 5 months, $\frac{3}{4}$ of the remainder in 10 months, and then had $500 left, what was the estate?

Ans. $3000.

Operation.— $1 - \frac{1}{3} = \frac{2}{3}$; $1 - \frac{3}{4} = \frac{1}{4}$

Then $\frac{1}{4} \times \frac{2}{3} = \frac{1}{6}$; $500 \div \frac{1}{6} =$

Ans. $3000.

After spending $\frac{1}{3}$ he has $\frac{2}{3}$ left. After spending $\frac{3}{4}$ of that, he must have $\frac{1}{4}$ of the $\frac{2}{3}$ left, which is $\frac{1}{6}$ of the whole; then 500 must be $\frac{1}{6}$ of the whole

6. From a drove of beeves I sell to A $\frac{1}{3}$, to B $\frac{3}{4}$ of the remainder, to C $\frac{2}{3}$ of the remainder, and have 50 beeves left; how many at first?   Ans. 450.

7. A cistern has two pipes for filling it,—one in 6 hours and the other in 12 hours; it also has a discharging pipe, which empties it in 5 hours. Now leave all open, in what time will the cistern fill?   Ans. 20 hours.

Operation. — $\frac{1}{6} + \frac{1}{12} = \frac{1}{4}$; $\frac{1}{4} - \frac{1}{5} = \frac{1}{20}$.

$1 \div \frac{1}{20} = 20$, Ans.

It is plain that the two filling pipes will fill $\frac{1}{4}$ of the cistern in 1 hour, while the discharging pipe will empty $\frac{1}{5}$ in 1 hour. Their difference shows *how much* is filled in 1 hour. Then 1, the whole cistern, divided by $\frac{1}{20}$ will give the hours.

8. A water tank has two filling pipes. The first would fill it in 40 minutes, the second in 50. It has a discharging pipe, which empties it in 25 minutes. Now suppose all turned on at once, how long would the tank be in filling?

Ans. 3 hours and 20 minutes.

9. John can do a job of work in 5 days, and James in 8 days; in what time can they both do the work together? Ans. $3\frac{3}{13}$ days.

Operation.—$8 \times 5 = 40$; $5 + 8 = 13$; $40 \div 13 = 3\frac{1}{13}$, Ans.

We may consider the work divided into 40 equal parts (or any other common multiple of 5 and 8), and we plainly see that John, working 5 days, will do *eight* of those parts each day, and James *five;* both together will do in 1 day 13 parts; hence it will require as many days as 13 is contained times in 40.

10. Sallie can make a coat in 6 days, Jane in 4 days; how long required for both to make it together? Ans. $2\frac{2}{5}$ days.

11. A can do a piece of work in 10 days, B in 12 days, and C in 15 days; in what time could all together accomplish it? Ans. 4 days.

12. Three men can do a job in 9 days. A alone can do it in 18 days, B in 27 days; in what time can C do it alone? Ans. 54 days.

Operation.—$27 - 18 = 9$. $27 \times 18 \div 9 = 54$, Ans.

In this case we divide any common multiple of the two given numbers by their difference, and the quotient shows the other number.

13. A man and his wife together can drink a demijohn of whisky in 12 days, but when the "old man" is absent it lasts the old lady 30 days. Now if the old lady leaves home, how long will it last the old man? Ans. 20 days.

14. Which will enclose the most ground,
A fence made square, or one made round,
Two panels to each rod of land,
Ten rails in each, we understand;
And every rail in each suppose
To just one acre of land enclose;
The next thing is to tell exact
How many acres in each tract?
Ans. $\begin{cases} 1024000 \text{ in the square.} \\ 804571\frac{3}{7} \text{ in the circle.} \end{cases}$

Operation.—$\frac{1}{160}$ = area of 1 rod square, which takes 80 rails to fence it.

Then $\frac{1}{160} : 80 : : 80 = 1024000$ rails or acres, Ans.

And as one side of the square equals the diameter of the circle, we multiply 1024000 by $\frac{11}{14}$ and obtain the circle = $804571\frac{3}{7}$ rails or acres, Ans.

15. A, B and C agree to grade a piece of railroad. A and B can do the work in 16 days, B and C in $13\frac{1}{3}$ days, and A and C in $11\frac{3}{7}$ days. In how many days can all do it, working together, and in how many days can each do it alone?

Operation.

$\frac{1}{16} = \frac{5}{80}$, what A and B do in 1 day.
$\frac{3}{40} = \frac{6}{80}$, what B and C do in 1 day.
$\frac{7}{80} = \frac{7}{80}$, what A and C do in 1 day.
$\frac{5}{80} + \frac{6}{80} + \frac{7}{80} = \frac{18}{80}$ what all three do in 2 days.
$\frac{18}{80} \div 2 = \frac{9}{80}$, what all will do in 1 day.
$1 \div \frac{9}{80} = 8\frac{8}{9}$ days, time A, B, and C will do the whole work together.
$\frac{9}{80} - \frac{5}{80} = \frac{4}{80}$ C alone.
$\frac{9}{80} - \frac{6}{80} = \frac{3}{80}$ A alone.
$\frac{9}{80} - \frac{7}{80} = \frac{2}{80}$ B alone.

And

$1 \div \frac{4}{80} = 20$ days for C.
$1 \div \frac{3}{80} = 26\frac{2}{3}$ days for A.
$1 \div \frac{2}{80} = 40$ days for B.

Analysis.—Since A and B can do the work in 16 days, they can do $\frac{1}{16}$ of it in 1 day; B and C, in $13\frac{1}{3}$ or $\frac{40}{3}$ days, they can do $\frac{3}{40}$ of it in 1 day; A and C in $11\frac{3}{7}$ or $\frac{80}{7}$ days, they can do $\frac{7}{80}$ in 1 day. Then A, B, and C, in two days can do $\frac{18}{80}$ of the work, and in 1 day $\frac{9}{80}$; and it will take them as many days working together to do the whole work as $\frac{9}{80}$ is contained times in 1, or $8\frac{8}{9}$ days. Now, if we take what any two do in 1 day from what three do in 1 day, the remainder will show what part the third does in 1 day. We thus find that A does $\frac{3}{80}$, B $\frac{2}{80}$, and C $\frac{4}{80}$.

Next denote the work by 1 and divide it by each of these fractions, and the quotient will express the days required by each to do it by himself.

16. A and B can do a piece of work, in 5 $\frac{5}{11}$ days, B and C in 6$\frac{2}{3}$, A and C in 6 days; in what time would all do the work together, and each alone? Ans. All, 4 days; A, 10; B, 12; C, 15.

WONDERS OF NUMERATION.

By the tables, as given in most of our school arithmetics, numeration is carried only to six places, or quadrillions, running up by terms derived from the Latin numerals.

A series of units of that extent would be beyond the power of man to comprehend, or even imagine. Even millions convey a very indefinite idea, and when it rises to billions, the mind can no longer grasp the number; for, though we may *read* the expression, it is very much as we read sentences in an *unknown language*. But we may perhaps assist the mind of the student by some little calculation. Often do we see *millions* spoken of in our national expenditures; and yet even that is an exceedingly large number, for if a man were to count fifteen hundred dollars an hour, and work faithfully eight hours a day, it would take him nearly three months to count a *million* of dollars; and if the dollars were silver, 1$\frac{3}{8}$ inches in diameter, laid touching each other in a straight line, they would reach over twenty-five miles, and thirty-one wagons, hauling two thousand pounds each would not be sufficient to haul them; and were they greenback dollar bills laid in this line, they would form a line over one hundred and ten miles long.

Our only plan, then, to understand this, is to group the number by imagining one thousand piles or lots, and one thousand dollars in each pile, when we can gain as distinct an idea of the number of piles as of the individuals of each pile. But suppose we extend even this mode to such a sum as our present na-

tional *debt*, and we are lost in wonder and amazement, and the mind is utterly bewildered. The present national debt is about three *billions*, and there are about two thousand clerks in the Treasury Department. Now, were they all to turn their attention to counting this money, and work eight hours a day counting one every second, it would require between three and four months to accomplish the work.

Dr. Thompson, Professor of Mathematics, at Belfast, Ireland, very justly remarks: "Such is the facility with which large numbers are expressed, both by figures and in language, that we generally have a very limited and inadequate conception of their real magnitude. The following considerations may perhaps assist in enlarging the ideas of the pupil on this subject:—

"To count a *million*, at one per section, would require between twenty-three and twenty-four days, of twelve hours each.

"The seconds in six thousand years are less than one-fifth of a *trillion*. A *quadrillion* of leaves of paper, each the two-hundredth part of an inch in thickness, would form a pile the height of which would be three hundred and twenty times the distance of the moon from the earth. Let it also be remembered that a *million* is equal to a thousand repeated a thousand times, and a *billion* equal to a million repeated a thousand times."

A rifle ball flies twelve hundred feet per second, and if one were fired at the moment one of the presidents of the United States takes his seat, and continued unabated for the *four years*, it would not travel *three millions* of miles. Suppose a man were to count one every second of time, day and night, without stopping to rest, eat, drink, or sleep, it would take him thirty-two years to count a *billion*, and thirty-two thousand years to count a *trillion*. What a limited idea we generally entertain of the immensity of numbers!

THE MILLER'S RULE FOR WEIGHING WHEAT.

Wheat weighing 58 pounds and upwards per bushel is considered merchantable wheat, and 60 pounds of merchantable wheat make a standard bushel. Hence, wheat weighing less than 60 pounds per bushel will lose in making up; but, weighing more, it will gain.

When wheat weighs less than 58 pounds per bushel, it is customary, on account of the inferior yield of light wheat, to take two pounds for one in making up the weight; hence, it will take 63 pounds to make up a bushel, provided the wheat weighs but 57, and 64 if the wheat weighs but 56 pounds per bushel.

CASE I.—To change merchantable wheat to standard weight.

RULE.—Bring the whole quantity of wheat to pounds and divide by 60.

EXAMPLE 1.—How many standard bushels of wheat are in 150 bushels, each weighing 58 pounds?

```
150     Or, each bushel lacks 2 lbs.;    150
 58                                        2
————                                      ——
1200                                   6,0)30,0
 750                                       ——
 ——     From 150 bush.    Deficiency,     5
6,0)870,0  Take   5                       ——
```

Ans. 145 b. Leaves 145, the answer.

2. How many standard bushels of wheat are in 80 bushels 45 pounds, weighing 63?

```
Bush.  lbs.
 80     45        Or, 80 bush.
 63                       3
————                    ——
285              6,0)24,0 excess of weight.
480                     ——
 ——                    4 bush.
6,0)508,5              30    45 lbs.
```

Ans. 84 b. 45 lbs.=84¾ b. Ans. 84 b. 45 lbs. or 3p.

3. How many standard bushels of wheat are in 175 bushels 37 pounds weighing 59? Ans. 172 bush. 42. lbs.

4. How many standard bushels are in 100 bushels 15 pounds, weighing 62 pounds per bushel? Ans. 103 bus. 35 lbs.

CASE II.—When wheat weighs less than 58.

RULE.—Bring the whole quantity to pounds, and divide by as many pounds as make a standard bushel of such wheat.

EXAMPLE 1.—How many bushels of good wheat are equal to 100 bushels weighing 57?

```
100        Or, 6 lbs. per bus.=600 lbs.
 57        63)600(9 bus. 33 lbs. defect.
 ——            567.
63)5700(90 bush. 30 lbs. ——
 567. From 100 bush.     33
 —— Take                9 33
30 lbs.                 ——
                   Ans. 90 30
```

NOTE.—The odd pounds in the above and following results are also subject to a small drawback, viz., 1 lb. in every 21 when the weight weighs 57; 1 in 16 when it weighs 56, and so on; consequently, the above ought, in strictness, to be 90 bushels, and rather more than 28½ pounds, but millers seldom make this deduction.

2. How many standard bushels of merchantable wheat will be equal to 250 bushels 18 lbs. weighing 56 lbs. per bushel? Ans. 219 bush. 2 lbs.

How much good wheat is equal to 1000 bushels weighing 55? Ans. 846 bush. 10 lbs.

NOTE.—before dismissing this rule it appears proper that a few remarks should be made, in order to show the young farmer the importance of understanding it properly. There are different methods of "making up wheat" (i. e., finding its merchantable value), and these methods give different results; hence the necessity of the subject being understood by all concerned. I shall not undertake to determine between the farmer and the miller which is or which is not the fair way; but, after explaining the principle, leave them to make their bargains as they may choose.

If I have a bushel of wheat that weighs but 57 pounds, then six pounds of the *same kind of wheat* will be necessary to make this a merchantable bushel, so that 63 pounds of this quality of wheat will make a standard bushel; and it is upon this supposition that the preceding calculations are founded. But a number of millers use a method of calculation by which they take *good* wheat for the odd pounds—*i. e.*, they take a bushel full, say 57 pounds, of the wheat they are measuring, and instead of taking six pounds more of *the same kind* to make it up, they take six pounds of good or *merchantable wheat*. Their method of calculation is as follows:—

Required, the good wheat in 1000 bushels weighing 55 ?

Defect, 10 lbs. per bush. 1000
                                          10
From 1000 bush.
Take   166 40   6,0)1000,0

Ans. 833 20   Bush. 166 40

We see that this gives a result nearly 13 bushels more in the miller's favor than the former method; and this I know to be practiced by many.

There is another method sometimes used by those who are not very scrupulous in their distinctions between right and wrong. They find the whole defect in pounds, and divide by the weight of a bushel of the wheat to find how many bushels of *that kind* of wheat will make up the defect, thus:—

Required, the good wheat in 1000 bushels weighing 55 pounds per bushel?

Defect, 10 lbs. per bushel=1000 in all.

From 1000 bush.   55)10000(181 bush. 45 lbs.
Take   181 45   55

Ans. 818 10   450
                             440
                             ———
                             100
                             55
                             ——
                             45 lbs.

We see that this method gives 15 bushels more to the miller than the last, and 28 more than the first. It, however, shows how much of the same kind of wheat must be added to the 1000 to make 1000 bushels of good wheat; viz., 181 bushels 45 pounds, for 1181 bushels 45 pounds, weighing 55, will just make 1000 bushels "made up weight." This method would be as erroneous as calculating discount by the rule for interest.

Another method is to take two pounds for one up to 58, and pound for pound afterwards. To do this bring the whole quantity to pounds, and if the wheat weigh 57 divide by 61; if 56, by 62, and so on. This appears more reasonable than the others, as it makes less difference between wheat barely merchantable and that which is not quite so. By the former rule, if the wheat weigh 58, two pounds per bushel will make it up, but if 57, six pounds are necessary; by this rule only one extra pound would be taken. In subtracting the odd pounds, the lower number being greatest (suppose the wheat to weigh, say 57), the calculator may be at a loss whether to take from 57, 60 or 63. In this case, let the running weight be what it may, we should take from the weight made up, as in example 1, case 2, we took from 63.

This subject is of importance to both farmers and millers, and if they do not attend to it they deserve to be cheated.

A TABLE FOR MEASURING TIMBER.

Quarter Girt.	Area.	Quarter Girt.	Area.	Quarter Girt.	Area.
Inches.	Feet.	Inches.	Feet.	Inches.	Feet.
6	.250	12	1.000	18	2.250
6¼	.272	12¼	1.042	18½	2.376
6½	.294	12½	1.085	19	2.506
6¾	.317	12¾	1.129	19½	2.640
7	.340	13	1.174	20	2.777
7¼	.364	13¼	1.219	20½	2.917
7½	.390	13½	1.265	21	3.062
7¾	.417	13¾	1.313	21½	3.209
8	.444	14	1.361	22	3.362
8¼	.472	14¼	1.410	22½	3.516
8½	.501	14½	1.460	23	3.673
8¾	.531	14¾	1.511	23½	3.835
9	.562	15	1.562	24	4.000
9¼	.594	15¼	1.615	24½	4.168
9½	.626	15½	1.668	25	4.340
9¾	.659	15¾	1.722	25½	4.516
10	.694	16	1.777	26	4.694
10¼	.730	16¼	1.833	26½	4.876
10½	.766	16½	1.890	27	5.062
10¾	.803	16¾	1.948	27½	5.252
11	.840	17	2.006	28	5.444
11¼	.878	17¼	2.066	28½	5.640
11½	.918	17½	2.126	29	5.840
11¾	.959	17¾	2.186	29½	6.044
				30	6.250

RULE.—(BY THE CARPENTERS' RULE.) Measure the circumference of the piece of timber in the middle and take a quar-of it in inches; call this the girt. Then set 12 on D to the length in feet on c, and against the girt in inches on D you will find the content in feet on c.

EXAMPLE.—If a piece of round timber be 18 feet long, and the quarter girt 24 inches, how many feet of timber are contained therein?

24 quarter girt.
24
———
96
48
———
576 square.
18
———
4608
576
———
144) 10368 (72 feet.
1008
———
288
288

Against 24 stands   4.00
Length,   18

Product,   72.00
Ans. 72 feet.

By the Carpenters' Rule.—12 on D: 18 on c: 24 on D: 72 on c.

PROBLEM 1.—To find the solid contents of squared or four-sided timber by the Carpenters' Rule—as 12 on D: length on c: quarter girt on D: solidity on c.

RULE 1.—*Multiply the breadth in the middle by the depth in the middle, and that product by the length, for the solidity.*

NOTE.—If the tree taper regularly from one end to the other, half the sum of the breadths of the two ends will be the breadth in the middle, and half the sum of the depths of the two ends will be the depth in the middle.

RULE II.—*Multiply the sum of the breadths of the two ends by the sum of the depths, to which add the product of the breadth and depth of each end; one sixth of this sum multiplied by the length will give the correct solidity of any piece of squared timber tapering regularly.*

PROBLEM II.—To find how much in length will make a solid foot, or any other assigned quantity of squared timber, of equal dimensions from end to end.

RULE.—*Divide 1728, the solid inches in a foot, or the solidity to be cut off, by the area of the end in inches, and the quotient will be the length in inches.*

NOTE.—To answer the purpose of the above rule, some carpenters' rules have a little table upon them, in the following form, called a *table of timber measure.*

0	0	0	0	9	0	11	3	9	inches.
144	36	16	9	5	4	3	2	1	feet.
1	2	3	4	5	6	7	8	9	side of the square.

This table shows that if the side of the square be one inch, the length must be 144 feet; if two inches be the side of the square, the length must be 36 feet, to make a solid foot.

PROBLEM III.—To find the solidity of round or unsquared timber.

RULE 1.—*Gird the timber round the middle with a string; one-fourth part of this girt squared and multiplied by the length will give the solidity.*

NOTE.—If the circumference be taken in inches and the length in feet, divide the last product by 144.

RULE II.—(BY THE TABLE.)—*Multiply the area corresponding to the quarter girt in inches, by the length of the piece of timber in feet, and the product will be the solidity.*

NOTE.—If the quarter girt exceed the table, take half of it, and four times the content thus formed will be the answer.

How do you do when the timber tapers?

Gird the timber at as many points as may be necessary, and divide the sum of the girts by their number for the mean girt, of which take one-fifth and proceed as before.

If a tree, girting 14 feet at the thicker end and 2 feet at the smaller end, be 24 feet in length, how many solid feet will it contain? Ans. 122.88.

A tree girts at five different places as follows: in the first, 9.43 feet, in the second 7.92 feet, in the third, 6.15 feet, in the fourth, 4.74 feet, and in the fifth, 3.16 feet; now, if the length of the tree be 17.25 feet, what is its solidity? Ans. 54.42499 cubic feet.

28

OF LOGS FOR SAWING.

What is often necessary for lumber merchants?

It is often necessary for lumber merchants to ascertain the number of feet of boards which can be cut from a given log, or, in other words, to find how many logs will be necessary to make a given amount of boards.

What is a standard board?

A standard board is one which is 12 inches wide; one inch thick, and 12 feet long; hence, a standard board is one inch thick and contains 12 square feet.

What is a standard saw log?

A standard log is 12 feet long and 24 inches in diameter.

How will you find the number of feet of boards which can be sawed from a standard log?

If we saw off, say two inches from each side, the log will be reduced to a square 20 inches on a side. Now, since a standard board is one inch in thickness, and since the saw cuts about one-quarter of an inch each time it goes through, it follows that one-fourth of the log will be consumed by the saw. Hence, we shall have $20 \times \frac{3}{4} =$ the number of boards cut from the log. Now, if the width of a board in inches be divided by 12, and the quotient be multiplied by the length in feet, the product will be the number of square feet in the board. Hence, $\frac{20}{12} \times$ length of log in feet = the square feet in each board. Therefore, $20 \times \frac{3}{4} \times \frac{20}{12} \times$ length of log = the square feet in all the boards = $20 \times 10 \times \frac{3}{4} \times \frac{2}{12} \times$ length of log = $20 \times 10 \times \frac{1}{8} \times$ length. And the same may be shown for a log of any length.

What, then, is the rule for finding the number of feet of boards which can be cut from any log whatever?

From the diameter of the log in inches

subtract four for the slabs; then multiply the remainder by half itself, and the product by the length of the log in feet, and divide the result by eight; the quotient will be the number of square feet.

EXAMPLE 1.—What is the number of feet of boards which can be cut from a standard log?

Diameter,	24 inches.
For slabs,	4
Remainder,	20
Half remainder,	10
	200
Length of log,	12
	8)2400
	300 = the number of feet.

2. How many feet can be cut from a log 12 inches in diameter and 12 feet long? Ans. 48.

3. How many feet can be cut from a log 20 inches in diameter and 16 feet long? Ans. 256.

4. How many feet can be cut from a log 24 inches in diameter and 16 feet long? Ans. 400.

5. How many feet can be cut from a log 28 inches in diameter and 14 feet long? Ans. 504.

CARPENTERS' AND JOINERS' WORK.

In what does Carpenters' and Joiners' work consist?

Carpenters' and joiners' work is that of flooring, roofing, etc., and is generally measured by the square of 100 square feet.

When is a roof said to have a true pitch?

In carpentry a roof is said to have a true pitch when the length of the rafters is three-fourths the breadth of the building. The rafters then are nearly at right angles. It is, therefore, customary to take once and a half times the area of the flat of the building for the area of the roof.

EXAMPLE. 1. How many squares of 100 square feet each, in a floor 48 feet 6 inches long and 24 feet 3 inches broad? Ans. 11 and 76$\frac{1}{2}$ sq. ft.

2. A floor is 36 feet 6 inches long and 16 feet 6 inches broad, how many squares does it contain? Ans. 5 and 98$\frac{1}{2}$ sq. ft.

3. How many squares are there in a partition 91 feet 9 inches long and 11 feet 3 inches high? Ans. 10 and 32 sq. ft.

If a house measure within the walls 52 feet 8 inches in length and 30 feet 6 inches in breadth, and the roof be of the true pitch, what will the roofing cost at $1.40 per square? Ans. $33.733.

OF BINS FOR GRAIN.

What is a bin?

It is a wooden box used by farmers for the storage of their grain.

Of what are bins generally made?

Their bottoms or bases are generally rectangles and horizontal, and their sides vertical.

How many cubic feet are there in a bushel?

Since a bushel contains 2150.4 cubic inches, and a cubic foot 1728 inches, it follows that a bushel contains 1$\frac{1}{4}$ cubic feet, nearly.

Having any number of bushels, how then will you find the corresponding number of cubic feet.

Increase the number of bushels one fourth itself, and the result will be the number of cubic feet.

EXAMPLE 1.—A bin contains 372 bushels; how many cubic feet does it contain?

$372 \div 4 = 93$; hence, $372 \times 93 = 465$ cubic feet.

2. In a bin containing 400 bushels how many cubic feet? Ans. 500.

How will you find the number of bushels which a bin of a given size will hold?

Find the content of the bin in cubic feet; then diminish the content by one fifth, and the result will be the content in bushels.

3. A bin is 8 feet long, 4 feet wide and 5 feet high; how many bushels will it hold?

$$8 \times 4 \times 5 = 160$$
then, $160 \div 5 = 32$: $160 - 32 = 128$ bushels = capacity of bin.

4. How many bushels will a bin contain which is 7 feet long, 3 feet wide and 6 feet in height. Ans. 100.8 bush.

How will you find the dimensions of a bin which shall contain a given number of bushels?

Increase the number of bushels one fourth itself, and the result will show the number of cubic feet which the bin will contain. Then, when the two dimensions of the bin are known, divide the last result by their product, and the quotient will be the other dimension.

5. what must be the height of a bin that will contain 600 bushels, its length being 8 feet and its breadth 4?

$600 \div 4 = 150$; hence, $600 + 150 = 750 =$ the cubic feet, and $8 \times 4 = 32$, the product of the given dimensions. Then, $750 \div 32 = 23.44$ feet, the height of the bin.

6. What must be the width of a bin that shall contain 900 bushels, the height being 12 and the length 10 feet?

$900 \div 4 = 225$; hence, $900 \times 225 = 1135 =$ the cubic feet; and $12 \times 10 = 120$ the product of the given dimensions. Then, $1155 \div 120 = 9.375$ feet, the width of the bin.

7. The length of a bin is 4 feet, its breadth 5 feet 6 inches, what must be its height that it may contain 136 bushels? Ans. 7 ft. 8 in.+

8. The depth of a bin is 6 feet 2 inches, the breadth 4 feet 8 inches; what must be the length that it may contain 200 bushels? Ans. 104 in.+

SLATERS' AND TILERS' WORK.

How is the content of a roof found?

In this work the content of the roof is found by multiplying the length of the ridge by the girt from eave to eave. Allowances, however, must be made for the double rows of slate at the bottom.

EXAMPLE 1.—The length of a slated roof is 45 feet 9 inches, and its girt 34 feet 3 inches; what is its content? Ans. 1566.9375 sq. ft.

2. What will the tiling of a barn cost at $3.40 per square of 100 feet, the length being 43 feet 10 inches and breadth 27 feet 5 inches on the flat, the eave board projecting 16 inches on each side and the roof being of the true pitch? Ans. $65.26.

BRICKLAYERS' WORK.

In how many ways is artificers' work computed?

Artificers' work in general is computed by three different measures; viz.:

1st. The linear measure, or, as it is called by mechanics, running measure.

2d. Superficial or square measure, in which the computation is made by the square foot, square yard, or by the square containing 100 square feet or yards.

3d. By the cubic or solid measure, when it is estimated by the cubic foot or the cubic yard. The work, however, is often estimated in square measure, and the materials for construction in cubic measure.

What proportion do the dimensions of a brick bear to each other?

The dimensions of a brick generally bears the following proportions to each other, viz.:—

Length=twice the width, and Width=twice the thickness; and hence the length is equal to four times the thickness.

What are the common dimensions of a brick? How many cubic inches does it contain?

The common length of a brick is 8 inches, in which case the width is 4 inches and the thickness 2 inches. A brick of this size contains 8×4×2=64 cubic inches; and since a cubic foot contains 1728 cubic inches, we have 1728÷64=27 the number of bricks in a cubic foot.

If a brick is 9 inches long, what will be its width and what its content?

If the brick is 9 inches long, then the width is 4½ inches, and the thickness 2¼; and then each brick will contain 9×4½×2¼=61⅛ cubic inches; and 1728÷91⅛=19 nearly, the number of bricks in a cubic foot. In the examples which follow we shall suppose the brick to be 8 inches long.

How do you find the number of bricks required to build a wall of given dimensions?

1st. Find the content of the wall in cubic feet.

2d. Multiply the number of cubic feet by the number of bricks, in a cubic foot, and the result will be the number of bricks required.

EXAMPLE 1. How many bricks, of 8 inches in length, will be required to build a wall 30 feet long, a brick and a half thick and 15 feet in height? Ans. 12150.

2. How many bricks, of the usual size, will be required to build a wall 50 feet long, two bricks thick, and 36 feet in height? Ans. 64800.

What allowance is made for the thickness of the mortar.

The thickness of mortar between the courses is nearly a quarter of an inch, so that four courses will give nearly one inch in height. The mortar, therefore, adds nearly one eighth to the height; but as one eighth is rather too large an allowance, we need not consider the mortar, which goes to increase the length of the wall.

3. How many bricks would be required in the first and second examples, if we make the proper allowance for mortar? Ans. 1st. 10631¼. 2d. 56700.

How do bricklayers generally estimate their work?

Bricklayers generally estimate their work at so much per thousand bricks.

What is the cost of a wall 60 feet long, 20 feet high and 2½ bricks thick, at $7.50 per thousand—which price we suppose to include the cost of the mortar?

If we suppose the mortar to occupy a space equal to one eighth the height of the wall, we must find the quantity of bricks under the supposition that the wall was 17½ feet in height. Ans. $354.37½.

In estimating the bricks for a house what allowances are made?

In estimating the bricks for a house, allowance must be made for the windows and doors.

OF CISTERNS.

What are cisterns?

Cisterns are large reservoirs constructed to hold water; and, to be permanent, should be made either of brick or masonry. It frequently occurs that they are to be so constructed as to hold given quantities of water, and then it becomes a useful and practical problem to calculate their exact dimensions.

How many cubic inches in a hogshead?

A hogshead contains 63 gallons, and a gallon contains 231 cubic inches; hence, $231 \times 63 = 14553$, the number of cubic inches in a hogshead.

How do you find the number of hogsheads which a cistern of given dimensions will contain?

1st. Find the solid content of the cistern in cubic inches.

2d. Divide the content so found by 14553 and the quotient will be the number of hogsheads.

EXAMPLE.—The diameter of a cistern is 6 feet 6 inches, and height 10 feet; how many hogsheads does it contain?

The dimensions reduced to inches are 78 and 120; then, the content in cubic inches, which is 573404.832, gives

$$573404.832 \div 14553 = 39.40 \text{ hogsheads}$$

nearly.

If the height of a cistern be given, how do you find the diameter, so that the cistern shall contain a given number of hogsheads?

1st. Reduce the height of the cistern to inches, and the content to cubic inches.

2d. Multiply the height by the decimal .7854.

3d. Divide the content by the last result and extract the square root of the quotient, which will be the diameter of the cistern in inches.

EXAMPLE.—The height of a cistern is 10 feet; what must be its diameter that it may contain 40 hogsheads? Ans. 78.6 in. nearly.

If the diameter of a cistern be given how do you find the height, so that the cistern shall contain a given number of hogsheads?

1st. Reduce the content to cubic inches.

2d. Reduce the diameter to inches, and then multiply its square by the decimal .7854.

3d. Divide the content by the last result, and the quotient will be the height in inches.

EXAMPLE. The diameter of a cistern is 8 feet; what must be its height that it may contain 150 hogsheads? Ans. 25 ft. 1 in., nearly.

MASONS' WORK.

What belongs to MASONRY, *and what measures are used?*

All sorts of stone work. The measure made use of is either superficial or solid.

Walls, columns, blocks of stone or marble are measured by the cubic foot; and pavements, slabs, chimney pieces etc., are measured by the square or superficial foot. Cubic or solid measure is always used for the materials, and the square measure is sometimes used for the workmanship.

EXAMPLE 1.—Required the solid content of a wall 53 feet 6 inches long, 12 feet 3 inches high and 2 feet thick. Ans. 1310¼ ft.

2. What is the solid content of a wall the length of which is 24 feet 3 inches, height 10 feet 9 inches, and thickness 2 feet? Ans. 521.375 ft.

3. In a chimney-piece we find the following dimensions:

Length of the mantel and slab 4 feet 2 inches.

Breadth of both together, 3 ft. 2 inches.

Length of each jam, 4 " 4 "

Breadth of both, 1 " 9 "

Required, the superficial content. Ans. 31 ft. 10′.

PLASTERERS' WORK.

How many kinds of plasterers' work are there, and how are they measured?

Plasterers' work is of two kinds, viz: ceiling, which is plastering on laths; and rendering, which is plastering on walls. These are measured separately.

The contents are estimated either by the square foot, the square yard, or the square of 100 feet.

Inriched mouldings, etc., are rated by the running or lineal measure.

In estimating plastering, deductions are made for chimneys, doors, windows, etc.

EXAMPLE 1. — How many square yards are contained in a ceiling 43 feet 3 inches long and 25 feet 6 inches broad? Ans. 122½ nearly.

2. What is the cost of ceiling a room 21 feet 8 inches by 14 feet 10 inches, at 18 cents per square yard? Ans. $6.42¼.

3. The length of a room is 14 feet 5 inches, breadth 13 feet 2 inches, and height to the under side of the cornice 9 feet 3 inches. The cornice girts 8½ inches, and projects 5 inches from the wall on the upper part next the ceiling, deducting only for one door 7 feet by 4; what will be the amount of the plastering?

Ans. { 53 yds. 5 ft. 3′ 6″ of rendering.
{ 18 yds. 5 ft. 6′ 4″ of ceiling.
{ 37 ft. 10′ 9″ of cornice.

How is the area of the cornice found in the above examples?

The mean length of the cornice both

in the length and breadth of the house is found by taking the middle line of the cornice. Now, since the cornice projects 5 inches at the ceiling, it will project $2\frac{1}{2}$ inches at the middle line; and, therefore, the length of the middle line along the length of the room will be 14 feet, and across the room 12 feet 9 inches. Then multiply the double of each of these numbers by the girth, which is $8\frac{1}{2}$ inches, and the sum of the products will be the area of the cornice.

PAINTERS' WORK.

How is painters' work computed, and what allowances are made?

Painters' work is computed in square yards. Every part is measured where the color lies, and the measuring line is carried into all the mouldings and cornices.

Windows are generally done at so much apiece. It is usual to allow double measure for carved mouldings, etc.

EXAMPLE 1.—How many yards of painting in a room which is 65 feet 6 inches in perimeter and 12 feet 4 inches in height? Ans. $89\frac{11}{14}$ sq. yds.

2. The length of a room is 20 feet, its breadth 14 feet 6 inches and height 10 feet 4 inches; how many yards of painting are in it—deducting a fireplace of 4 feet by 4 feet 4 inches, and two windows, each 6 feet by 3 feet 2 inches. Ans. $73\frac{2}{7}$ sq. yds.

PAVERS' WORK.

How is pavers' work estimated?

Pavers' work is done by the square yard, and the content is found by multiplying the length and breadth together.

EXAMPLE 1.—What is the cost of paving a sidewalk, the length of which is 35 feet 4 inches and breadth 8 feet 3 inches, at 54 cents per square yard? Ans. $17.48 9.

2. What will be the cost of paving a rectangular court yard, whose length is 63 feet and breadth 45 feet, at 2s. 6d. per square yard, there being, however, a walk running lengthwise 5 feet 3 inches broad, which is to be flagged with stone costing 3s per square yard? Ans. £40 5s. $10\frac{1}{2}d$

PLUMBERS' WORK.

Plumbers' work is rated at so much a pound, or else by the hundred weight. Sheet lead, used for gutters, etc., weighs from 6 to 12 pounds per square foot. Leaden pipes vary in weight according to the diameter of their bore and thickness.

The following table shows the weight of a square foot of sheet lead, according to its thickness; and the common weight of a yard of leaden pipe, according to the diameter of the bore:—

Thickness of Lead.	Pounds to a Square Foot.	Bore of Leaden Pipes.	Pounds per Yard.
Inch. $\frac{1}{16}$	5.899	Inch. $0\frac{3}{4}$	10
$\frac{1}{8}$	6.554	1	12
$\frac{1}{8}$	7.373	$1\frac{1}{4}$	16
$\frac{1}{4}$	8.427	$1\frac{1}{2}$	18
$\frac{1}{4}$	9.831	$1\frac{3}{4}$	21
$\frac{1}{3}$	11.797	2	24

EXAMPLE 1.—What weight of lead of $\frac{1}{12}$ of an inch in thickness will cover a flat 15 feet 6 inches long, and 10 feet 3 inches broad, estimating the weight at 6 lbs. per square foot? Ans. 8 cwt. 2 qr. 1¼ lb.

2. What will be the cost of 130 yards of leaden pipe of an inch and a half bore, at 8 cents per pound, supposing each yard to weigh 18 lbs? Ans. $187.20.

3. The lead used for a gutter is 12 feet 5 inches long and 1 foot 3 inches broad, what is its weight, supposing it to be $\frac{1}{8}$ of an inch in thickness? Ans. 101 lbs. 12 oz. 13.6 dr.

4. What is the weight of 96 yards of leaden pipe of an inch and a quarter bore. Ans. 13 cwt. 2 qr. 24 lbs.

5. What will be the cost of a sheet of lead 16 feet 6 inches long and 10 feet 4 inches broad, at 5 cents per pound, the lead being $\frac{1}{8}$ of an inch in thickness? Ans. 83.81.

WEIGHTS AND MEASURES.

TROY WEIGHT.

By this weight gold, silver, platina, and precious stones (except diamonds) are estimated.

20 mites........1 grain.
20 grains ...1 pennyw't.
20 pennywt's....1 ounce
12 ounces.......1 pound.

Any quantity of gold is supposed to be divided into 24 parts, called *carats*. If pure, it is said to be 24 carats fine; if there is 22 parts of pure gold and 2 parts of alloy it is said to be 22 carats fine. The standard of American coin is nine-tenths pure gold, and is worth $20.67. What is called the *new standard*, used for watch cases, etc., is 18 carats fine.

The term carat is also applied to a weight of 3½ grains troy, used in weighing diamonds; it is divided into 4 parts, called *grains;* 4 grains troy are thus equal to 5 grains diamond weight.

APOTHECARIES' WEIGHT, USED IN MEDICAL PRESCRIPTIONS.

The pound and ounce of this weight are the same as the pound and ounce troy, but differently divided.

20 grains troy. ......1 scruple.
3 scruples.... ......1 drachm.
8 drachms...... ...1 ounce troy.
12 ounces............1 pound troy.

Druggists *buy* their goods by

AVOIRDUPOIS WEIGHT.

By this weight all goods are sold except those named under troy weight.

27$\frac{1}{32}$ grains........1 drachm.
16 drachms... ....1 ounce.
16 ounces.........1 pound.
28 pounds. ....1 quarter.
4 quarters........1 hundred weight.
20 hundred weight..1 ton.

The grain avoirdupois, though never used, is the same as the grain in troy weight. 7,000 grains make the avoirdupois pound, and 5,760 grains the troy pound. Therefore, the troy pound is less than the avoirdupois pound in the proportion of 14 to 17, nearly; but the troy ounce is greater than the avoirdupois ounce in the proportion of 79 to 72, nearly. In times past it was the custom to allow 112 pounds for a hundred weight, but usage, as well as the laws of a majority of the States, at the present time call 100 pounds a hundred weight.

APOTHECARIES' FLUID MEASURE.

60 minims.... .1 fluid drachm.
8 fluid drachms..1 ounce (troy).
16 ounces (troy)...1 pint.
8 pints..........1 gallon.

35

MEASURE OF CAPACITY FOR ALL LIQUORS.

5 ounces avoirdupois, of water make 1 gill.

4 gills.........1 pint = 34⅔ cubic inches, nearly.

2 pints.........1 quart = 69⅓ cubic inches.

4 quarts.......1 gallon = 277¼ do. inches.

31½ gallons........1 barrel.

42 gallons........ 1 tierce.

63 gals., or 2 bbls...1 hogshead

2 hogsheads......1 pipe or butt.

2 pipes..........1 ton.

The gallon must contain exactly 10 pounds avoirdupois of pure water at a temperature of 62°, the barometer being at 30 inches. It is the standard unit of measure of capacity for liquids and dry goods of every description, and is ⅛ larger than the old wine measure, ₃¹₂ larger than the old dry measure, and ₆¹₀ less than the old ale measure. The wine gallon must contain 231 cubic inches.

MEASURE OF CAPACITY FOR ALL DRY GOODS.

4 gills.............1 pint = 34½ cub. in., nearly.
2 pints...,........1 quart = 69½ cubic inches.
4 quarts...........1 gallon = 277¼ cubic inches.
2 gallons..........1 peck = 554½ cubic inches.
4 pecks, or 8 gals..1 bushel = 2150½ cubic inches.
8 bushels..........1 quarter = 10¼ cubic ft., nearly.

When selling the following articles a barrel weighs as here stated:—

For rice, 600 lbs.; flour, 196 lbs.; powder, 25 lbs.; corn, as bought and sold in Kentucky, Tennessee, etc., 5 bushels of shelled corn; as bought and sold at New Orleans, a flour barrel full of ears; potatoes, as sold in New York, a barrel contains 2¾ bushels; pork, a barrel is 200 lbs., distinguished in quality by "clear," "mess," "prime;" a barrel of beef is the same weight.

The legal bushel of America is the old Winchester measure of 2,150.42 cubic inches. The imperial bushel of England is 2,218.142 cubic inches, so

that 32 English bushels are about equal to 33 of ours.

Although we are all the time talking about the price of grain, etc., by the bushel, we sell by weight, as follows:—

Wheat, beans, potatoes, and clover-seed, 60 lbs. to the bushel; corn, rye, flax-seed and onions, 56 lbs.; corn on the cob, 70 lbs.; buckwheat, 52 lbs.; barley, 48 lbs.; hemp seed, 44 lbs.; timothy seed, 45 lbs.; castor beans, 46 lbs.; oats, 35 lbs.; bran, 20 lbs.; blue grass seed, 14 lbs.; salt, the real weight of coarse salt is 85 lbs.; dried apples, 24 lbs.; dried peaches, 33 lbs.; according to some rules, but others are 22 lbs. per bushel, while in Indiana dried apples and peaches are sold by the heaping bushel; so are potatoes, turnips, onions, apples, etc., and in some sections oats are heaped. A bushel of corn in the ear is three heaped half bushels, or four even full.

In Tennessee a hundred ears of corn is sometimes counted as a bushel.

A hoop 18½ inches diameter 8 inches deep holds a Winchester bushel. A box 12 inches square, 7 and 7₃¹₀ deep, will hold half a bushel. A heaping bushel is 2,815 cubic inches.

CLOTH MEASURE.

2½ inches........1 nail.

4 nails.........1 quarter of a yard.

4 quarters......1 yard.

FOREIGN CLOTH MEASURE.

2½ quarters... ....1 ell Hamburg.

3 quarters........1 ell Flemish.

5 quarters........1 ell English.

6 quarters........1 ell French.

MEASURE OF LENGTH.

12 inches ......1 foot.

3 feet..........1 yard.

5½ yards... ...1 rod, pole or perch.

40 poles ......1 furlong.

8 furlongs, or 1760 yds. 1 mile.

69₁⁶₀ miles..... } 1 degree of a great circle of the earth.

By scientific persons and revenue officers the inch is divided into tenths, hundredths, etc. Among mechanics the inch is divided into eights. The division of the inch into 12 parts, called lines, is not now in use.

A standard English mile, which is the measure that we use, is 5,280 feet in length, 1,760 yards, or 320 rods. A strip one rod wide and one mile long is two acres; By this it is easy to calculate the quantity of land taken up by roads, and also how much is wasted by fences.

GUNTER'S CHAIN — USED FOR LAND MEASURE.

$7\frac{92}{100}$ inches..............1 link.
100 links, or 66 feet, or 4 poles, 1 chain.
10 chains long by 1 broad, or
    10 sq. chains...........1 acre.
80 chains...................1 mile.

SURFACE MEASURE.

144 square inches.....1 square foot.
  9 square feet........1 square yard.
$30\frac{1}{4}$ square yards......1 square rod or
                           perch.
40 square perches....1 rood.
  4 roods...........1 acre.
640 acres ....1 square mile.

Measure 209 feet on each side and you have a square acre within an inch.

The following gives the comparative size in square yards of acres in different countries:—

English acre, 4,840 square yards ; Scotch, 6,150 ; Irish, 7,840; Hamburg, 11,545 ; Amsterdam, 9,722 ; Dantzic, 6,650 ; France (hectare), 11,960 ; Prussia (morgen), 3,053.

This difference should be borne in mind in reading of the products per acre in different countries. Our land measure is that of England.

GOVERNMENT LAND MEASURE.

A township—36 sections, each a mile square

A section—640 acres.

A quarter section, half a mile square —160 acres.

An eighth section, half a mile long, north and south, and a quarter of a mile wide—80 acres.

A sixteenth section, a quarter of a mile square—40 acres.

The sections are all numbered 1 to 36, commencing at the north-east corner thus :—

6	5	4	3	2	$\frac{N\ E}{S\ E}$ s w
7	8	9	10	11	12
18	17	16*	15	14	13
19	20	21	22	23	24
30	29	28	27	26	25
31	32	33	34	35	36

The sections are divided into quarters, which are named by the cardinal points, as in section 1. The quarters are divided in the same way. The description of a forty-acre lot would read: The south half of the west half of the south-west quarter of section 1 in township 24, north of range 7 west, or as the case might be; and sometimes will fall short and sometimes overrun the number of acres it is supposed to contain.

SQUARE MEASURE.

FOR CARPENTERS, MASONS, ETC.

144 square inches.....1 square foot.
9 sq. ft.. or 1,296
    sq. in...........1 square yard.
100 square feet........1 sq. of flooring.
                     roofing, etc.
$30\frac{1}{4}$ square yards... .1 square rod.
36 square yards......1 rood of buid'g,

*School section.

GEOGRAPHICAL OR NAUTICAL MEASURE.

6 feet...............1 fathom.
110 fathoms, or 660 ft. .1 furlong.
6075 ⅞ feet............1 nautical mile.
3 nautical miles......1 league.
20 leag's, or 60 geo. ms. 1 degree.
360 degrees............The earth's
    circum.=24,855½ miles, nearly.
The nautical mile is 795⅞ feet longer
than the common mile.

MEASURE OF SOLIDITY.

1728 cubic inches....1 cubic foot.
27 cubic feet......1 cubic yard.
16 cubic feet.... .1 cord ft., or ft.
           of wood.
8 cord ft., or 128
    cub. ft.........1 cord.
40 ft. of round, or 50
    ft. of hewn timber 1 ton.
42 cubic feet......1 ton of shipping.

ANGULAR MEASURE, OR DIVISIONS OF THE
CIRCLE.

60 seconds.........1 minute.
60 minutes..........1 degree.
30 degrees..........1 sign.
90 degrees.........1 quadrant.
360 degrees.........1 circumference.

MEASURE OF TIME.

60 seconds......... 1 minute.
60 minutes..........1 hour.
24 hours............1 day.
7 days............1 week.
28 days...........1 lunar month.
28, 29, 30 or 31 days. .1 calen'r month.
12 calendar months...1 year.
365 days............1 com. year.
366 days............1 leap year.
365¼ days ....1 Julian year.
365 d. 5 h. 48 m. 49 s., 1 Solar year.
365 d. 6 h. 9 m. 12 s...1 Siderial year.

ROPES AND CABLES.

6 feet..............1 fathom.
120 feet..............1 cable length.

MISCELLANEOUS IMPORTANT FACTS
ABOUT WEIGHTS AND MEASURES.

BOARD MEASURE.

Boards are sold by superficial measure
at so much per foot of one inch or less
in thickness, adding one fourth to the
price for each quarter inch thickness
over an inch.

GRAIN MEASURE IN BULK.

Multiply the width and length of the
pile together, and that product by the
height, and divide by 2,150, and you
have the contents in bushels.

If you wish the contents of a pile of
ears of corn, or roots, in heaped bushels,
ascertain the cubic inches and divide by
2,818.

CAPACITY OF CISTERNS OR WELLS.

Tabular view of the number of gallons contained in the clear between the
brick work for each ten inches of depth :

DIAMETER.		GALLONS.
2 feet equal..................		19
2½ " "		30
3 " "		44
3½ " "		60
4 " "		78
4½ " "		97
5 " "		122
5½ " "		148
6 " "		176
6½ " "		207
7 " "		240
7½ " "		275
8 " "		313
8½ " "		353
9 " "		396
9½ " "		461
10 " "		489
11 " "		592
12 " "		705
13 " "		827
14 " "		959
15 " "		1101
20 " "		1958
25 " "		3059

TO MEASURE CORN IN THE CRIB.

Corn is generally put up in cribs made of rails, but the rule will apply to a crib of any size or kind.

Two cubic feet of good, sound, dry corn in the ear will make a bushel of shelled corn. To get, then, the quantity of shelled corn in a crib of corn in the ear, measure the length, breadth, and height of the crib, *inside of the rail;* multiply the length by the breadth, and the product by the height; then divide the result by two, and you have the number of bushels of shelled corn in the crib.

In measuring the height, of course the height of the corn is intended. And there will be found to be a difference in measuring corn in this mode, between fall and spring, because it shrinks very much in the winter and spring, and settles down.

TO GAUGE A BOX OR GRANARY.

RULE.—*Take the dimensions in feet; multiply them together, and that product by* $\frac{4}{5}$, *and you have the bushels of shelled grain. Or,*

Place the dimensions on the right with 4 under them, and five on the left as a divisor.

NOTE.—This rule is based upon the Winchester bushel, 2150.4 cubic inches. If we divide the number of cubic inches in one foot—1728, by 2150, we have a quotient of 8, or $\frac{8}{10}=\frac{4}{5}$, therefore, a cubic foot holds $\frac{4}{5}$ of a cubic bushel; hence, taking $\frac{4}{5}$ of a cubic feet of any space will give us *bushels* of shelled grain.

1. How many bushels of wheat will a box hold that is 5 feet long, 4 feet wide, and 3 feet deep?

Ans. 48 bush.

Operation.

	5
5	4
	3
	4
	—
	48 Ans.

2. I have a granary 10 feet long, 8 feet high, and 7 feet wide; how many bushels will it hold?

Ans. 448.

3. If A. B. Nuckolls' granary, 12 feet long, 9 feet wide, 7 feet high, is just half full, how many bushels of wheat has he? Ans. 302$\frac{2}{5}$ bus.

4. A miller has a "toll box" 6 feet long, 5 feet wide, 3 feet deep; how much corn would it hold?

Ans. 72 bus.

5. If a box is 2 feet in the clear every way, how much corn would it hold?

Ans. 6$\frac{2}{5}$ bus.

6. I have a granary 12$\frac{1}{2}$ feet long, 8$\frac{3}{4}$ feet wide, 4$\frac{1}{5}$ feet deep; how many bushels will it hold? Ans. 367$\frac{1}{2}$ bus.

TO GAUGE A CRIB.

RULE.—*Take the dimensions in feet; multiply them together and that product by* $\frac{2}{5}$, *and the result will be bushels. Or,*

Place the dimensions on the right with 2 under them, and 5 on the left as a divisor.

NOTE.—Taking $\frac{2}{5}$ of the cubic feet gives the bushels of corn in the ear, that is the number of bushels the "bulk" would "shell out." Corn is supposed to shell out half the bulk, hence we take $\frac{2}{5}$ instead of $\frac{4}{5}$, as in the preceding rule. The truth is, there can be no exact rule to gauge a crib, as different kinds of corn will "shell out" differently, and a large crib will shell out more than a small one in proportion to size, as the *pressure* is much greater, and will cause it to gauge *less* than it will shell out.

However, the rule above is the nearest correct of any ever published.

1. How many bushels of corn will a crib contain whose dimensions are 15 feet long, ten feet high, 8 feet wide?
Ans. 480 bushels.

2. How many barrels of corn are there in a crib 20 feet long, 12 feet high, 10 feet wide? Ans. 192 bbls.

3. How many barrels of corn will a crib contain whose dimensions are 20 feet long, 12 feet high, 6 feet wide?
Ans. 115½ bbls.

4. I have a crib 15¾ feet long, 7¼ feet high, and 6⅞ feet wide; what are the contents in barrels? Ans. 57¾ bbls.

Note.—When inches are given, consider them fractions of a foot.

5. Stephen Cantrell has a crib 15 feet 4 inches long, 10 feet 6 inches wide, 8 feet 7 inches high; how many barrels will it hold? Ans. 110+bbls.

GAUGING CASKS.

Rule. — *Take the distance in inches from the centre of the bung inside, diagonally, to the chine; cube it, and divide by 370, and the quotient will express the gallons. Should there be a remainder, multiply by 4, and continue the division for quarts, by 2, for pints, etc.*

Note.—If the bung is not in the centre, measure both ways to chine; add the two results together, and take half the sum; then proceed as above.

This standard number 370 is derived from actual experiment. The measurement of a regular shaped cask cubed as above, divided by the actual capacity by the English gallon pot, gave the standard 370. So we may take it and divide it into the cube of any cask, and we have the capacity.

1. How many gallons will a hogshead hold measuring 37 inches from the centre of the bung inside to the chine?
Ans. 136 gals. 3 qts. 1 pt.

Operation.

$37 \times 37 \times 37 = 50653 \div 370 = 136$ gallons.

1st remainder $333 \times 4 = 1332 \div 370 = 3$ quarts.

2d remainder $222 \times 2 = 444 \div 370 = 1$ pint.

2. A cask measures 16 inches from the centre of the bung, diagonally, to the chine; what is its capacity?
Ans. 11 gals. 2 gills.

3. A cask measures 18 inches, diagonally, to the chine inside, one way, and 19 inches the other, what will it hold?
Ans. 17 gals. and 3+gills.

4. I have a small cask measuring 13 inches to the chine inside; what does it hold? Ans. 5 gals. 3 qts. 1 pt. 2 gills.

MENTAL OPERATIONS IN FRACTIONS.

To square any number containing ½ as 6½, 9½.

Rule.—*Multiply the whole number by the next higher whole number and annex ¼ to the product.*

Example 1. What is the square of 7½?
Ans. 56¼.

. We simply say 7 times 8 are 56, to which we add ¼.

2. What will 9½ lbs. beef cost at 9½ cts. a lb.?

3. What will 12½ yds. tape cost at 12½ cts. a yd.?

4. What will 5½ lbs. nails cost at 5½ cts. a lb.?

5. What will $11\frac{1}{2}$ yds. tape cost at $11\frac{1}{2}$ cts a yd.?

6. What will $19\frac{1}{2}$ bu. bran cost at $19\frac{1}{2}$ cts. a bu.?

Reason.—We multiply the whole number by the next higher whole number, because half of any number taken twice and added to its square is the same as to multiply the given number by *one* more than itself. The same principle will multiply any two *like* numbers together, when the sum of the fractions is *one*, as $8\frac{1}{3}$ by $8\frac{2}{3}$, or $11\frac{3}{8}$ by $11\frac{5}{8}$, etc. It is obvious that, to multiply any number by any two fractions whose sum is *one*, the sum of the products *must be the original number*, and adding the number to its square is simply to multiply it by *one* more than itself,—for instance, to multiply $7\frac{1}{4}$ by $7\frac{3}{4}$ we simply say 7 times 8 are 56, and then, to complete the multiplication, we add, of course, the product of the fractions ($\frac{3}{4}$ times $\frac{1}{4}$ are $\frac{3}{16}$), making $56\frac{3}{16}$ the answer.

To square any number containing $\frac{1}{2}$.

Rule.—*Multiply the whole number by the next higher whole number and annex $\frac{1}{4}$ to the product.*

11. What is the square of $8\frac{1}{2}$? Ans. $72\frac{1}{4}$.

We simply say 9 times 8 are 72 and annex $\frac{1}{4}$.

12. What will $12\frac{1}{2}$ pounds beef come to a $12\frac{1}{2}$ cents a pound? Ans. \$1.56$\frac{1}{4}$.

13. What will $6\frac{1}{2}$ pounds spike come to at $6\frac{1}{2}$ cents a pound? Ans. $42\frac{1}{4}$.

To multiply any two *like* numbers together when the sum of the fractions is one.

Rule.—*Multiply the whole number by the next higher whole number, and to the product add the product of the fractions.*

Remark.—To find the product of the fractions multiply the numerators together for a new numerator and the denominators for a new denominator.

14. Multiply $6\frac{2}{5}$ by $6\frac{3}{5}$. Ans. $42\frac{6}{25}$.

Explanation.—Multiply 6, the whole number, by 7, the next higher whole number $= 42$. We then multiply the numerators of the fraction, $2 \times 3 = 6$, and the denominators $5 \times 5 = 25$, making the product $\frac{6}{25}$, which we add to the product of the whole number, 42.

15. Multiply $7\frac{1}{3}$ by $7\frac{2}{3}$. Ans. $56\frac{2}{9}$.

16. Multiply $11\frac{3}{7}$ by $11\frac{4}{7}$. Ans. $132\frac{12}{49}$.

17. Multiply $29\frac{1}{3}$ by $29\frac{2}{3}$. Ans. $870\frac{2}{9}$,

To multiply any two like numbers together, each of which has a fraction with a like denominator, as $3\frac{3}{4}$ by $5\frac{1}{4}$, or $6\frac{3}{8}$ by $7\frac{7}{8}$, etc.

Rule.—*Add to the multiplicand the fraction of the multiplier and multiply this sum by the whole number; to the product add the product of the fractions.*

18. Multiply $6\frac{1}{4}$ by $5\frac{3}{4}$, Ans. $35\frac{3}{16}$.

The sum of $6\frac{1}{4}$ and $\frac{3}{4}$ is 7, so we simply say 5 times 7 are 35; to this we add the product of the fractions, $\frac{3}{4}$ times $\frac{1}{4}$ are $\frac{3}{16} = 35\frac{3}{16}$ Ans.

19. Multiply $9\frac{1}{4}$ by $8\frac{2}{4}$. Ans. $78\frac{2}{16}$.

The sum of $9\frac{1}{4}$ and $\frac{2}{4}$ is $9\frac{3}{4}$, and 8 times $9\frac{3}{4}$ are 78, to which add the product of the fractions.

WHERE THE SUM OF THE FRACTIONS IS ONE.

To multiply any two numbers whose difference is *one* and the sum of the fractions is one.

Rule.—*Multiply the larger number, increased by one, by the smaller number; then square the fraction of the larger number, and subtract its square from one.*

PRACTICAL EXAMPLES FOR BUSINESS MEN.

1. What will $9\frac{1}{4}$ lbs. sugar cost at $8\frac{3}{4}$ cts. per pound?

41

Here we multiply 9, increased by 1, by 8, thus: 8×10 are 80, and set down the result; then from 1 we subtract the square of $\frac{1}{4}$ thus: $\frac{1}{4}$ squared is $\frac{1}{16}$, and 1 less $\frac{1}{16}$ is $\frac{15}{16}$.

$9\frac{1}{4}$
$8\frac{3}{4}$
——
$80\frac{15}{16}$

2. What will $8\frac{2}{3}$ bu. coal cost at $7\frac{1}{3}$ cts. a bu. ?

Here we multiply 8, increased by 1, by 7, thus: 7 times 9 are 63, and set down the result; then from 1 we subtract the square of $\frac{2}{3}$, thus: $\frac{2}{3}$ squared is $\frac{4}{9}$, and 1 less $\frac{4}{9}$ is $\frac{5}{9}$

$8\frac{2}{3}$
$7\frac{1}{3}$
——
$63\frac{5}{9}$

3. What will $11\frac{2}{13}$ bu. seed cost at $\$10\frac{11}{13}$ a bu. ?

Here we multiply 11, increased by 1, by 10, thus: 10 times 12 are 120, and set down the result; then from 1 we subtract the square of $120\frac{22}{169}$, thus: $\frac{2}{13}$ squared is $\frac{4}{169}$, and 1 less $\frac{4}{169}$ is $\frac{165}{169}$.

$11\frac{2}{13}$
$10\frac{11}{13}$
——
$120\frac{22}{169}$

4. How many square inches in a floor $99\frac{2}{3}$ inches wide and $98\frac{1}{3}$ in. long ? Ans. $9800\frac{2}{9}$.

METHOD OF OPERATION.

EXAMPLE FIRST.

Multiply $6\frac{1}{4}$ by $6\frac{1}{4}$ in a single line.

Here we add $6\frac{1}{4}+\frac{1}{4}$, which gives $6\frac{1}{2}$; this multiplied by the 6 in the multiplier, $6 \times 6\frac{1}{2}$ gives 39, to which we add the product of the fractions; thus $\frac{1}{4} \times \frac{1}{4}$ gives $\frac{1}{16}$, added to 39 completes the product.

$6\frac{1}{4}$
$6\frac{1}{4}$
——
$39\frac{1}{16}$

EXAMPLE SECOND.

Multiply $11\frac{1}{4}$ by $11\frac{3}{4}$ in a single line.

Here we would add $11\frac{1}{4}+\frac{3}{4}$, which gives 12; this multiplied by the 11 in the multiplier gives 132, to which we add the product of the fractions; thus $\frac{3}{4} \times \frac{1}{4}$ gives $\frac{3}{16}$, which added to 132 completes the product.

$11\frac{1}{4}$
$11\frac{3}{4}$
——
$132\frac{3}{16}$

EXAMPLE THIRD.

Multiply $12\frac{1}{2}$ by $12\frac{3}{4}$ in a single line.

Here we add $12\frac{1}{2}+\frac{3}{4}$, which gives $13\frac{1}{4}$; this multiplied by the 12 in the multiplier, $12 \times 13\frac{1}{4}$, gives 159, to which add the product of the fractions; thus $\frac{3}{4} \times \frac{1}{2}$ gives $\frac{3}{8}$, which added to 159 completes the product.

$12\frac{1}{2}$
$12\frac{3}{4}$
——
$159\frac{3}{8}$

WHERE THE SUM OF THE FRACTIONS IS ONE.

To multiply any two *like* numbers together when the sum of the fractions is one.

RULE.—*Multiply the whole number by the next higher whole number, after which add the product of the fractions.*

N. B.—In the following examples the product of the fractions are obtained *first*, for convenience:—

PRACTICAL EXAMPLES FOR BUSINESS MEN.

Multiply $3\frac{3}{4}$ by $3\frac{1}{4}$ in a single line.

Here we multiply $\frac{1}{4} \times \frac{3}{4}$, which gives $\frac{3}{16}$, and set down the result; then we multiply the 3 in the multiplicand, increased by unity, by the 3 in the multiplier, 3×4, which gives 12 and completes the product.

$3\frac{3}{4}$
$3\frac{1}{4}$
——
$12\frac{3}{16}$

Multiply $7\frac{2}{8}$ by $7\frac{3}{8}$ in a single line.

Here we multiply $\frac{3}{8} \times \frac{2}{8}$, which gives $\frac{6}{64}$, and set down the result: then we multiply the 7 in the multiplicand, increased by unity, by the 7 in the multiplier, 7×8, which gives 56, and completes the product.

$7\frac{2}{8}$
$7\frac{3}{8}$
——
$56\frac{6}{64}$

Multiply $11\frac{1}{3}$ by $11\frac{2}{3}$ in a single line.

Here we multiply $\frac{2}{3} \times \frac{1}{3}$, which gives $\frac{2}{9}$, and set down the result; then we multiply the 11 in the multiplicand, increased by unity, by the 11 in the multiplier, 11×12, which gives 132, and completes the product.

$11\frac{1}{3}$
$11\frac{2}{3}$
——
$132\frac{2}{9}$

EXAMPLE FOURTH.

Multiply $16\frac{2}{3}$ by $16\frac{1}{3}$ in a single line.

Here we multiply $\frac{1}{3} \times \frac{2}{3}$, which $16\frac{2}{3}$ gives $\frac{2}{9}$, and set down the result; $16\frac{1}{3}$ then we multiply the 16 in the mul- —— tiplicand, increased by unity, by $272\frac{2}{9}$ the 16 in the multiplier, 16×17, which gives 272, and completes the product.

EXAMPLE FIFTH.

Multiply $29\frac{1}{2}$ by $29\frac{1}{2}$ in a single line. Here we multiply $\frac{1}{2} \times \frac{1}{2}$, which $29\frac{1}{2}$ gives $\frac{1}{4}$, and set down the result; $29\frac{1}{2}$ then we multiply the 29 in the mul- —— tiplican, increased by unity, by $870\frac{1}{4}$ the 29 in the multiplier, 29×30, which gives 870, and completes the product.

NOTE.—The system of multiplication introduced in the preceding examples applies to all numbers. Where the sum of the fractions is *one*, and the whole numbers are alike, or differ by *one*, the learner is requested to study well these useful properties of numbers.

WHERE THE FRACTIONS HAVE A LIKE DE-NOMINATOR.

To multiply any two *like* numbers together, each of which has a fraction with a *like* denominator, as $4\frac{3}{8} \times 4\frac{5}{8}$, or $11\frac{1}{4} \times 11\frac{3}{4}$, or $10\frac{3}{8} \times \frac{5}{8}$, etc.

RULE. — *Add to the multiplicand the fraction of the multiplier, and multiply this sum by the whole number, after which add the product of the fractions.*

PRACTICAL EXAMPLES FOR BUSINESS MEN.

N. B.—In the following example the sum of the fractions is *one:*—
1. What will $9\frac{3}{4}$ lbs. of beef cost at $9\frac{1}{4}$ cts. a lb.?

The sum of $9\frac{3}{4}$ and $\frac{1}{4}$ is 10, so we $9\frac{3}{4}$ simply say 9 times 10 are 90; then $9\frac{1}{4}$ we add the product of the fractions, —— $\frac{1}{4}$ times $\frac{3}{4}$ are $\frac{3}{16}$. $90\frac{3}{16}$
N. B.—In the following example the sum of the fractions is *less* than *one:*

2. What will $8\frac{1}{4}$ yds. tape cost at $8\frac{3}{4}$ cts. a yd.?

The sum of $8\frac{1}{4}$ and $\frac{2}{4}$ is $8\frac{3}{4}$, so we $8\frac{1}{4}$ simply say 8 times $8\frac{3}{4}$ are 70; then $8\frac{3}{4}$ we add the product of the fractions, —— $\frac{2}{4}$ times $\frac{1}{4}$ are $\frac{2}{16}$, or $\frac{1}{8}$. $70\frac{1}{8}$

N. B.—In the following example the sum of the fraction is *greater* than *one:*—
3. What will $4\frac{3}{8}$ yds. cloth cost at $5\frac{7}{8}$ a yd.?

The sum of $4\frac{3}{8}$ and $\frac{7}{8}$ is $5\frac{1}{4}$, so we $4\frac{3}{8}$ simply say 4 times $5\frac{1}{4}$ are 21; then $4\frac{7}{8}$ we add the product of the fractions, —— $\frac{7}{8}$ times $\frac{3}{8}$ are $\frac{21}{64}$. $21\frac{21}{64}$

N. B.—Where the fractions have different denominators reduce them to a common denominator.

RAPID PROCESS FOR MULTIPLYING MIXED NUMBERS

A valuable and useful rule for the accountant in the practical calculations of the counting room.
To multiply any two numbers together, each of which involves the fraction $\frac{1}{2}$—as $7\frac{1}{2} \times 9$, etc.

RULE.—*To the product of the whole numbers add half their sum, plus $\frac{1}{4}$.*

EXAMPLES FOR MENTAL OPERATIONS.

1 What will $3\frac{1}{2}$ dozen eggs cost at $7\frac{1}{2}$ cts. a doz.?
Here the sum of 7 and 3 is 10, $3\frac{1}{2}$ and half this sum is 5, so we simply $7\frac{1}{2}$ say 7 times 3 are 21 and 5 are 26, —— to which we add $\frac{1}{4}$. $26\frac{1}{4}$

N. B.—If the sum be an odd number call it one less, to make it even, and in such cases the fraction must be $\frac{3}{4}$.

2. What will $11\frac{1}{2}$ lbs. cheese cost at $9\frac{1}{2}$ cts. a lb.?

3. What will $8\frac{1}{2}$ yds. tape cost at $15\frac{1}{2}$ cts. a yd.?

4. What will $7\frac{1}{2}$ lbs. rice cost at $13\frac{1}{2}$ cts. a lb.?

5 What will $10\frac{1}{2}$ bu. coal cost at $12\frac{1}{2}$ cts. a bu.?

REASON. — In explaining the above rule we add half their sum, because half of either number added to half the other would be half their sum, and we add $\frac{1}{4}$ because $\frac{1}{2}\times\frac{1}{2}$ is $\frac{1}{4}$. The same principle will multiply any two numbers together, each of which has the same fraction—for instance, if the fraction was $\frac{1}{3}$ we would add one-third their sum; if $\frac{1}{4}$, we would add three-fourths their sum, etc.; and then, to complete the multiplication, we would add, of course, the product of the fractions.

6. Multiply $4\frac{3}{8}$ by $4\frac{7}{8}$. Ans. $21\frac{21}{64}$.

The sum of $4\frac{3}{8}$ and $\frac{7}{8}$ is $5\frac{1}{4}$, and 4 times $5\frac{1}{4}$ is 21; add $\frac{3}{8}\times\frac{7}{8}=\frac{21}{64}$. $21\frac{21}{64}$ Ans.

To multiply any two numbers together, each of which involves the fraction $\frac{1}{2}$.

RULE.—*To the product of the whole numbers add half their sum, plus $\frac{1}{4}$*

7. Multiply $3\frac{1}{2}\times7\frac{1}{2}$. Ans. $26\frac{1}{4}$:

Solution.—The sum of 3 and 7 are 10, and one-half this sum is 5, so we say, 7 times 3 are 21 and 5 are 26, to which we annex $\frac{1}{4}$. $26\frac{1}{4}$ Ans.

8. What will $7\frac{1}{2}$ lbs. cheese cost at $13\frac{1}{2}$ cts. a lb.? Ans. $1.01\frac{1}{4}$.

REMARK.—If the sum be an odd number call it one less, to make it even; in which case the fraction must be $\frac{3}{4}$.

9. What will $8\frac{1}{2}$ lbs, of sugar cost at $15\frac{1}{2}$ cts. a lb.? Ans. $1.31\frac{3}{4}$.

Here, $8+15=23$, being an odd number, we make it one less, 22, one-half of which is 11. Then 8 times 15 are 120, and 11 are 131, to which we add $\frac{3}{4}$.

The same principle will multiply any two numbers together, each of which has the same fraction. For instance, if the fraction was $\frac{1}{5}$, we would add one-fifth their sum; if $\frac{3}{4}$, we would add three-fourths their sum; if $\frac{2}{3}$, add two-thirds their sum, etc., after which, of course, add the product of their fractions.

10. Multiply $8\frac{2}{3}\times7\frac{2}{3}$. Ans. $66\frac{4}{9}$.

The sum of 8 and 7 are 15, two-thirds of which is 10. We then say 8 times 7 are 56 and 10 makes 66, and add $\frac{2}{3}\times\frac{2}{3}$ $=\frac{4}{9}$.

INTEREST

Is a sum paid for the use of money.

Principal is a sum for the use of which interest is paid.

Amount is the sum of the principal and interest.

Rate per cent., commonly expressed decimally as hundredths, is the sum per cent. paid for the use of one dollar annually.

Simple Interest is the sum paid for the use of the principal only during the whole time of the loan.

Legal Interest is the rate per cent. established by law.

Usury is illegal interest, or a greater per cent. than the legal rate.

It is contended by many statesmen that the rate of interest should not be established by statute, but that money is only a commodity that, like every other article of traffic, should be governed by the law of supply and demand. If money is scarce the rate would be high; if plenty, then low. But as banks and other great moneyed institutions have the power, to a great extent, of controling the quantity of money in the market, thereby oppressing the great majority of the people, and taking advantage of the times of scarcity, public opinion, at least, has established the law of *usury*.

To find the interest if the time consists of years.

RULE.—*Multiply the principal by the rate per cent., and that product by the number of years.*

EXAMPLE 1.—What is the interest of $150 for 3 years, at 8 per cent.?

$$\begin{array}{r} \$150 \\ .08 \\ \hline 12.00 \\ 3 \\ \hline \$36.00 \text{ Ans.} \end{array}$$

The decimal for 8 per cent. is .08. There being two places of decimals in the multiplier we point off two places in the product.

To find the interest when the time consists of years and months.

RULE. — *Reduce the time to months. Multiply the principal by the rate per cent., divide the product by 12, and the quotient multiplied by the number of months will be the interest required.*

OR BY CANCELLATION.—*Place the principal, rate and time in months on the right of the line, and 12 on the left, then cancel.*

2. Find the interest of $240 for 2 years and 7 months, at 7 per cent.

$$\begin{array}{lr} \text{Principal,} & \$240 \\ \text{Rate,} & .07 \\ \hline & 12)16.80 \\ \hline & 1.40 \\ 2 \text{ yrs.} + 7 \text{ mos.} & 31 \\ \hline & 1.40 \\ & 4.20 \\ \hline & \$43.40 \text{ Ans.} \end{array}$$

BY CANCELLATION.

$$12 \left| \begin{array}{r} \$240 \quad 20 \\ 7 \\ 31 \end{array} \right.$$

$20 \times 7 \times 31 = \$43.40.$ Ans.

BANKERS' METHOD OF COMPUTING INTEREST AT 6 PER CENT. FOR ANY NUMBER OF DAYS.

RULE. — *Draw a perpendicular line, cutting off the two right-hand figures of the $, and you have the interest for 60 days_at 6 per cent.*

NOTE.—The figures on the left of the line are dollars, and those on the right are decimals of dollars.

EXAMPLE 1. What is the interest of $423 for 60 days, at 6 per cent.?

$423=the principal.
$4 | 23 cts.=interest for 60 days.

NOTE.—When the time is more or less than 60 days, first get the interest for 60 days, and from that to the time required.

EXAMPLE 2.—What is the interest of $124 for 15 days at 6 per cent.?

Days.   Days.
15=¼ of 60

$124=principal.
4) 1 | 24 cts.=interest for 60 days.

  | 31 cts.=interest for 15 days.

EXAMPLE 3.—What is the interest of $123.40 for 90 days, at 6 per cent.?

Days. Days. Days.
90 = 60 + 30

$123.40=principal.
2)1 | 2340=interest for 60 days.
    | 6170=interest for 30 days.

Ans $ | 851=interest for 90 days.

EXAMPLE 4.—What is the interest of $324 for 75 days, at 6 per cent.?

Days. Days. Days.
75 = 60 + 15

$324=principal.
4)3 | 24 cts. interest for 60 days.
    | 81 cts. interest for 15 days.

Ans. $4 | 05 cts. interest for 75 days.

REMARK.—This system of computing interest is very easy and simple, especially when the days are aliquot parts of 60, and one simple division will suffice. It is used extensively by a large majority of our most prominent bankers; and, indeed, is taught by most all commercial colleges as the shortest system of computing interest.

METHOD OF CALCULATING AT DIFFERENT PER CENTS.

This principle is not confined alone to 6 per cent., as many suppose who teach and use it. It is their custom *first* to find the interest at 6 per cent., and from that to other per cents.; but it is equally applicable for *all* per cents., from 1 to 15, inclusive.

The following table shows the different per cents., with the time that a given number of $ will amount to the same number of cents when placed at interest:—

RULE.—*Draw a perpendicular line, cutting off the two right-hand figures of $, and you have the interest at the following per cents.:—*

Interest at 4 per cent. for 90 days.
Interest at 5 per cent. for 72 days.
Interest at 6 per cent. for 60 days.
Interest at 7 per cent. for 52 days.
Interest at 8 per cent. for 45 days.
Interest at 9 per cent. for 40 days.
Interest at 10 per cent. for 36 days.
Interest at 12 per cent. for 30 days.
Interest at 7-30 per cent. for 50 days.
Interest at 5-20 per cent. for 70 days.
Interest at 10-40 per cent. for 35 days.
Interest at $7\frac{1}{2}$ per cent. for 48 days.
Interest at $4\frac{1}{2}$ per cent. for 80 days.

NOTE.—The figures on the left of the perpendicular line are dollars, and on the right decimals of dollars. If the dollars are less than 10 prefix a cipher.

EXAMPLE 1.—What is the interest of $120 for 15 days at 4 per cent.?

Days   Days.
$120=principal.   15=⅙ of 90
6)1 ⌈ 20 cts.=interest for 90 days.
    ⌊ 20 cts.=interest for 15 days.

EXAMPLE 2.—What is the interest of $132 for 13 days, at 7 per cent.?

Days.   Days.
$132=principal.   13=¼ of 52.
4)1 | 32 cts.=interest for 52 days.
    | 33 cts.=interest for 13 days.

EXAMPLE 3.—What is the interest of $520 for 9 days at 8 per cent.?

Days.   Days.
$520=principal.   9=⅕ of 45
5)5 | 20 cts.=interest for 45 days.
$1 | 04 cts.=interest for 9 days.

EXAMPLE 4.—What is the interest of $462 for 64 days, at $7\frac{1}{2}$ per cent.?

Days. Days. Days.
$462=principal.   64=48+16
3)4 | 62 cts.=interest for 48 days.
$1 | 54 cts.=interest for 16 days.
—————
$6 | 16 cts.=interest for 64 days.

REMARK.—We have now illustrated several examples by the different per cents., and if the student will study carefully the solution to the above examples, he will in a short time be very rapid in this mode of computing interest.

NOTE.—The preceding mode of computing interest is derived and deduced from the cancelling system, as the ingenious student will readily see. It is a short and easy way of finding interest for days, when the days are even or aliquot parts; but when they are not multiples, and three or four divisions are necessary, the cancelling system is much more simple and easy. We will here illustrate an example to show the difference.

Required, the interest of $420 for 49 days, at 6 per cent.:—

BANKERS' METHOD.   CANCELLING METHOD.

2)4|20 cts.=int. for 60 days. |420—70
——|——   6—36| 6
2)2|10 cts.=int. for 30 days. |49
5)1|05 cts.=int. for 15 days. | 70
3) |21 cts.=int. for 3 days.   ———
   | 7 cts.=int. for 1 day. $3.430 Ans.
——————
$3 | 43 cts.=int. for 49 days.

The cancelling method is much more brief; we simply cancel 6 in 36, and the quotient ·6 into 420; there is no devisor left; hence, 70×49 gives the interest at *once*.

If the time had been 15 or 20 days, the bankers' method would have been equally as short, because 15 and 20 are aliquot parts of 60. The superiority of the cancelling system above all others is this, it takes advantage of the *principal* as well as the *time*.

For the benefit of the student, and for the convenience of business men, we will investigate this system to its full extent, and explain how to take advantage of the *principal* when no advantage can be taken of the *days*. This is one of the most important characteristics of interest, and very often saves much labor. *It should be used when the days are not even or aliquot parts.*

The following table shows the different sums of money (at the different per cents.) that bear one cent interest a day; hence, the time in days is always the interest in cents; therefore, to find the interest on any of the following notes, at the per cent. attached to it in the table, we have the following

RULE. — *Draw a perpendicular line, cutting off the two right-hand figures of the days for cents, and you have the interest for the given time.*

Interest of $90 at 4 per cent. for 1 day is 1 cent.

Interest of $72 at 5 per cent. for 1 day is 1 cent.

Interest of $60 at 6 per cent. for 1 day is 1 cent.

Interest of $52 at 7 per cent. for 1 day is 1 cent.

Interest of $45 at 8 per cent. for 1 day is 1 cent.

Interest of $40 at 9 per cent. for 1 day is 1 cent.

Interest of $36 at 10 per cent. for 1 day is 1 cent.

Interest of $30 at 12 per cent. for 1 day is 1 cent.

Interest of $50 at 7.30 per cent. for 1 day is 1 cent.

Interest of $70 at 5.20 per cent. for 1 day is 1 cent.

Interest of $35 at 10.40 per cent. for 1 day is 1 cent.

Interest of $48 at $7\frac{1}{2}$ per cent. for 1 day is 1 cent.

Interest of $80 at $4\frac{1}{2}$ per cent. for 1 day is 1 cent.

Interest of $24 at 15 per cent. for 1 day is 1 cent.

NOTE.—The 7-30 Government Bonds are calculated on the base of 365 days to the year, and the 5-20s and 10-40s on the base of 364 days to the year.

PROBLEMS SOLVED BY BOTH METHODS.

We will now solve some examples by both methods to further illustrate this system, and for the purpose of teaching the pupil how to use his judgment. He will then have learned a rule *more valuable than all others.*

EXAMPLE 5.—What is the interest on $180 for 75 days, at 5 per cent. ?

Operation by taking advantage of the dollar.

75=the days. $60×3=$180.
$0 | 75 cts.=the int. of $60 for 75 days.
   | 3   Multiply by 3.

Ans. $2 | 25 cts.=the int. of $180 for 75 days.

Operation by the Bankers' method.

$180=the principal. 60 da.+15da.= 75 da.

4)$1 | 80 cts.=the int. for 60 days.
     | 45 cts.=the int. for 15 days.

Ans. $2 | 25 cts=the int. for 75 days.

By the first method we multiplied by 3, because 3×$60=$180. By the second method we added on ¼, because 60 da.+ ⁹⁰da.=75 da.

N. B.—When advantage can be taken of both time and principal, if the student wishes to prove his work he can first work it by the Bankers' method, and then by taking advantage of the principal, or *vice versa*. And as the two operations are entirely different, if the same result is obtained by each, he may fairly conclude that the work is correct.

PARTIAL PAYMENTS ON NOTES, BONDS AND MORTGAGES.

To compute interest on notes, bonds and mortgages, on which partial payments have been made, two or three rules are given. The following is called the common rule, and applies to cases where the time is short, and payments made within a year of each other. This rule is sanctioned by custom and *common law*; it is true to the principles of simple interest, and requires no special enactment. The other rules are rules of *law*, made to suit such cases as require (either expressed or implied) annual interest to be paid, and, of course, apply to no business transactions closed within a year.

RULE. — *Compute the interest of the principal sum for the whole time to the day of settlement, and find the amount. Compute the interest on the several payments from the time each was paid to the day of settlement; add the several payments and the interest on each together and call the sum the amount of the payments; subtracting the amount of the payments from the amount of the principal will leave the sum due.*

EXAMPLE.—A gave his note to B for $10,000; at the end of 4 months A paid $6,000, and at the expiration of another 4 months he paid an additional sum of $3,000; how much did he owe B at the close of the year?

```
Principal ................$10,000
Interest for the whole time...    600
                              ----------
Amount........... ..... $10,600
1st payment......$6,000.
Interest, 8 mos....   240.
2d payment....... 3,000
Interest, 4 mos ....   60
                  -------
Amount. ....... $9,300          9,300
                              ----------
             Due....$1,300
```

PROBLEMS IN INTEREST.

There are *four* parts or quantities connected with each operation in interest; these are the *Principal, Rate per cent., Time, Interest or Amount.*

If any *three* of them are given the *other* may be found.

Principal, interest and time given to and the rate per cent.

EXAMPLE 1.—At what rate per cent. must $500 be put on interest to gain $120 in 4 years?

OPERATION.

```
  $500
   .01
  -----
  5.00
     4
  -----
20.00)120.00(6 per cent.  Ans.
      120.00
```

BY ANALYSIS.

The interest of $1 for the given time at 1 per cent. is 4 cts. $500 will be 500 times as much=500×.04=$20. Then if $20 give 1 per ct., $120 will give $\frac{120}{20}$ =6 per cent.

RULE.—*Divide the given interest by the interest of the given sum at one per cent. for the given time, and the quotient will be the rate per cent. required.*

Principal, interest and rate per cent. given to find the time.

EXAMPLE 2.—How long must $500 be on interest at 6 per cent. to gain $120 ?

OPERATION.

$500
·.06

30.00) 120.00 (4 years. Ans.
120.00

ANALYSIS

We find the interest of $1 at the given rate for one year is six cents. $500 will be, therefore, 500 times as much $= 500 \times .06 = \$30.00$. Now, if it take one year to gain $30, it will require $\frac{120}{30}$ to gain $120=4$ years. Ans.

EQUATION OF PAYMENTS.

Equation of payments is the process of finding the equalized or average time for the payment of several sums due at different times, without loss to either party.

To find the average or mean time of payment when the several sums have the same date.

RULE.—*Multiply each payment by the time that must elapse before it becomes due; then divide the sum of these products by the sum of the payments, and the quotient will be the average time required.*

NOTE.—When a payment is to be made down it has no product, but it must be added with the other payments in finding the average time.

EXAMPLE.—I purchased goods to the amount of $1,200; $300 of which I am to pay in 4 months, $400 in 5 months, and $500 in 8 months. How long a

credit ought I to receive if I pay the whole sum at once? Ans. 6 months.

Mo.	Mo.	
4 X 300 =	1200	A credit on $300 for 4 months is the same as the credit on $1 for 1200 months.
5 X 400 =	2000	A credit on $400 for 5 months is the same as the credit on $1 for 2000 months.
8 X 500 =	4000	A credit on $500 for 8 months is the same as the credit on $1 for 4000 months.
1200)	7200(6 mo.	Therefore, I should have the same credit as the credit on $1 for 7200 months; and on $1200, the whole sum, one twelve hundredth part of 7200 months, which is 6 months.
	7200	

This rule is the one usually adopted by merchants, although not strictly correct; still, it is sufficiently accurate for all practical purposes.

To find the average or mean time of payment when the several sums have different dates.

EXAMPLE. — Purchased of James Brown; at sundry times and on various terms of credit, as by the statement annexed. When is the *medium* time of payment?

Jan. 1, a bill amounting to $360, on 3 months' credit.

Jan. 15, a bill amounting to $186, on 4 months' credit.

March 1, a bill amounting to $450, on 4 months' credit.

May 15, a bill amounting to $300, on 3 months' credit.

June 20, a bill amounting to $500, on 5 months' credit.

Ans. July 24, or in 115 days.

Due, April	1,	$360.		
" May	15,	186 X 44	=	8184
" July	1,	450 X 91	=	40950
" Aug.	15,	300 X 136	=	40800
" Nov.	20,	500 X 233	=	116500

1796÷into)206434(114$\frac{830}{808}$ dys

We first find the time when each of the bills will become due. Then, since it will shorten the operation and bring the same result, *we take the time when the first bill becomes due,* instead of its *date,* for the *period* from which to compute the average time. Now, since

April 1 is the period from which the average time is computed, no time will be reckoned on the first bill, but the time for the payment of the second bill extends 44 days beyond April 1, and we multiply it by 44.

Proceeding in the same manner with the remaining bills, we find the average time of payment to be 114 days and a fraction from April 1, or on the 24th of July.

RULE.—*Find the time when each of the sums become due, and multiply each sum by the number of days from the time of the earliest payment to the payment of each sum respectively; then proceed as in the last rule, and the quotient will be the average time required, in days, from the earliest payment.*

NOTE.—Nearly the same result may be obtained by reckoning the time in months.

In mercantile transactions it is customary to give a credit of from 3 to 9 months on bills of sale. Merchants, in settling such accounts as consist of various items of debit and credit for different times, generally employ the following

RULE.—*Place on the debtor or credit side such a sum (which may be called MERCHANDISE BALANCE) as will balance the account. Multiply the number of dollars in each entry by the number of days from the time the entry was made to the time of settlement, and the merchandise balance by the number of days for which credit was given. Then multiply the difference between the sum of the debit and the sum of the credit products by the interest of $1 for one day; this product will be the INTEREST BALANCE.*

When the sum of the debit products exceeds the sum of the credit products the interest balance is in favor of the debit side; but when the sum of the credit products exceeds the sum of the debit products it is in favor of the credit side. Now, to the merchandise balance add the interest balance, or substract it as the case may require, and you obtain the CASH BALANCE.

A has with B the following account.—

1849.			Dr.
Jan. 2.	To merchandise,		$200
April 20.	"	"	400
1849.			Cr.
Feb. 29.	By merchandise,		$100
May 10.	"	"	300

If interest is estimated at 7 per cent., and a credit of 60 days is allowed on the different sums, what is the cash balance August 20, 1849? Ans. $206.54.

EXPLANATION.—Without interest the cash balance would be $200.

The object of these changes is to give the learner an accurate and complete knowledge of numbers and of division, and the result is not the only object sought for, as many young learners suppose.

How many times is 75 contained in 575? or divide 575 by 75. Ans. $7\frac{2}{3}$.

Divide 800 by $12\frac{1}{2}$. Quotient, 64.

Divide 27 by $16\frac{2}{3}$. Quotient, $1\frac{62}{100}$, or $1\frac{31}{50}$.

A person spent $6 for oranges, at $6\frac{1}{4}$ cents a piece; how many did he purchase? Ans. 96.

When two or more numbers are to be multiplied together, and one or more of them have a cipher on the right, as 24 by 20, we may take the cipher from one number and annex it to the other without affecting the product: thus 24×20 is the same as 240×2; $286 \times 1300 = 28600 \times 13$; and $350 \times 70 \times 40 = 35 \times 7 \times 4 \times 1000$, etc.

Every fact of this kind, though extremely simple, will be very useful to those who wish to be skillful in operation.

NOTE.—If there are ciphers at the | neglected to the close of the opera-
right hand either of the multiplier or | tion, when they must be annexed to the
multiplicand, or of both, they may be | product.

TABLE FOR BANKING AND EQUATION.

*Showing the number of Days from any date in one Month to the same date in any
other Month.   Example: How many days from the 2d of February to the
2d of August ?   Look for February at the left hand, and August at the top—
in the angle is 181.   In leap year, add 1 day if February be included.*

From To	Jan.	Feb.	March	April.	May.	June.	July.	Aug.	Sept.	Oct.	Nov.	Dec.
January	365	31	59	90	120	151	181	212	243	273	304	334
February	334	365	28	59	89	120	150	181	212	242	273	303
March	306	337	365	31	61	92	122	153	184	214	245	275
April...	275	306	334	365	30	61	91	122	153	183	214	244
May...	245	276	304	335	365	31	61	92	123	153	184	214
June	214	245	273	304	334	365	30	61	92	122	153	183
July.............	184	215	243	274	304	335	365	31	62	92	123	153
August............	153	184	212	243	273	304	334	365	31	61	92	122
September....	122	153	181	212	242	273	303	334	365	30	61	91
October	92	123	151	182	212	243	273	304	335	365	31	61
November........	61	92	120	151	181	212	242	273	304	334	365	30
December...	31	62	90	121	151	182	212	243	274	304	335	365

www.ingramcontent.com/pod-product-compliance
Lightning Source LLC
Chambersburg PA
CBHW022024190326
41519CB00010B/1583